PRAISE FOR

Hollywood Jock

"Rob Ryder is an immensely talented writer. This book is an insider's look at Hollywood—a world few people truly understand."

—Beau Bridges, Emmy and Golden Globe–winning actor

"*In Living Color* meets *Get Shorty* on the hardwood of the Staples Center!"

—Will Staeger, executive producer of ESPN original entertainment and critically acclaimed author of *Public Enemy*

"Part *Entourage*, part *Sporting News*, part *Daily Variety*, and part *Family Circus*. A fantastic, funny, and gut-wrenching look at getting a sports movie to the screen. Somehow Rob Ryder gives you everything."

—Travis Rodgers, producer of *The Jim Rome Show*

"This book is fantastic. It brings together the two things I love the most: movies and sports."

—John Cheng, head of feature development, Rat Entertainment

© Andrea Gleysteen

About the Author

A technical consultant on many sports-themed movies, Rob Ryder played basketball at Princeton and wrote the column "Hollywood Jock" for ESPN.com. Also a screenwriter, Ryder has spent too many years in development hell.

HOLLYWOOD JOCK

365 Days, Four Screenplays, Three TV Pitches, Two Kids, and One Wife Who's Ready to Pull the Plug

HARPER

NEW YORK • LONDON • TORONTO • SYDNEY

HOLLYWOOD JOCK

Rob Ryder

HARPER

HarperCollins books may be purchased for educational, business, or sales promotional use. For information please write: Special Markets Department, HarperCollins Publishers, 10 East 53rd Street, New York, NY 10022.

FIRST EDITION

Designed by Justin Dodd

Library of Congress Cataloging-in-Publication Data
Ryder, Rob.
Hollywood jock : 365 days, four screenplays, three pitches, two kids, and one wife who's ready to pull the plug / Rob Ryder.—1st ed.
p. cm.
ISBN-10: 0-06-079150-0
ISBN-13: 978-0-06-079150-6
1. Ryder, Rob. 2. Screenwriters—United States—Biography. 3. Hollywood (Los Angeles, Calif.)—Biography. 4. Motion picture industry—United States. I. Title.

PS3618.Y36H65 2006
812'.6—dc22
[B] 2005044798

06 07 08 09 10 ❖/RRD 10 9 8 7 6 5 4 3 2 1

For Andrea

HOLLYWOOD JOCK

INTRODUCTION

Sports will rip your heart out. Dash your hopes, shatter your confidence, and leave you bleeding. Sports will offer up just enough victories to make all those defeats truly bitter. It's a world of violence and intimidation. Vitriolic coaches, heartless fans, and cynical critics.

Same goes for the movie business.

So what career path did I choose, lamebrain knucklehead that I am? The path of most resistance.

The path of the Hollywood Jock. And this is my story.

It starts at the beginning of the end.

Twelve months ago, I made a pact with my wife. "Honey, give me one more year. If I don't make a sizable score, I'll walk away from Hollywood for good."

"Promise?" she asked.

"Promise," I said. "One year, that's all I'm asking for."

"Okay, one year."

"I love you," I said.

"So does the dog," she answered.

I immediately went down into my basement office to take stock. I've been in the movie business since 1975, starting back in New

York. I worked as a production assistant, as a locations scout, and as an assistant director, all the while writing spec screenplays. I moved to Hollywood and optioned a script, then sold another. I got my first agent. I got my first studio writing assignment. I changed agents. The development deals started piling up, one after another, year after year, but nothing got produced. Then I was hired by my friend and mentor, Ron Shelton, to help coordinate the basketball sequences for *White Men Can't Jump*. This led to a second career as a sports adviser that nicely complemented my writing gigs. I sold another script. And another. I wrote and directed a Showtime 30 minute movie (the only project I've seen produced). But I still couldn't get a feature to the big screen.

The writing gigs began to dry up. I turned to more sports coordinating jobs—features, TV shows, commercials. A dream job in many ways—working with movie stars, famous coaches, and pro athlete Hall-of-Famers—but it gets old, asking millionaires to stop hanging on the rims.

So I took a step sideways, spending several years trying to launch a four-on-four full-court summer basketball league. My partners and I raised some money and got oh-so-close (this dream is still alive), but we were ahead of our time. The money dried up. I considered my options, and they weren't pretty. Teach school in the inner city with an emergency credential. Swing a hammer on a framing crew. Write advertising copy for Dunkin' Donuts. Or beg my wife for one more year to revive my last-gasp Hollywood career. Thank God for women who keep the faith.

With the one-year reprieve, I rolled up my sleeves and got to work, and right out of the box, I caught a break.

I knew a guy at ESPN.com, the 800-pound gorilla of sports Web sites. I had an idea for a weekly column, "Hollywood Jock." They bought it. Five hundred bucks a week. Barely enough to put a ding in our monthly nut, but a start. An auspicious beginning to a harrowing year. So here we go, and to paraphrase Bette Davis, "Fasten your seat belts. We're in for a bumpy ride."

1

WELCOME TO L.A.

Pookey's late. But so am I, striding along Melrose Avenue through the great saucy mix of hipster Los Angelenos—every size, color, and flavor—tattooed and pierced, the young women showing all that skin between their ya-yas and their lowslung jeans. I fall in behind one—a long artistic tat running across the girl's back right above her ass crack. I slow my pace and stay behind her for a block. Grrrrwl.

I find the entrance to Pookey's office, 7551 Melrose, pound up the ratty stairs and stick my head into a threadbare office. A beautiful, exotic, tawny-skinned, long-limbed creature sits sorting press releases at a desk. I can just imagine the tat across her lower back.

"Are you Maya?"

"No. Are you?"

"No." I stare at her. She stares back with big black serious eyes.

"I must be in the wrong place," I say.

"Again?" she asks.

"Yeah, again," I say, then add, "My whole life."

She suddenly smiles wide, revealing a great set of white teeth and a glint of braces. "I'm Rasha," she says. "She's Maya."

I step inside the office and spot a second beautiful exotic woman. A couple years older perhaps. A couple inches shorter. Straight black hair, copper skin. She's talking into a headset, typing on a laptop. She's wearing a white shirt with that one extra button undone that can make a man's day. She's got a cast on her foot. She looks over and sees my eyes move from her buttons to her cast.

"Wild sex," she says by way of explanation, and I know I'm in the right place. Look, people have to work for a living—we all know that. But it takes a guy like Pookey to understand let's at least put some juice in it. Spice it up a bit. Rasha turns out to be Egyptian, Maya, Indonesian. Welcome to L.A.

Pookey's got it going on. African-American, five foot three, literally, and one of the best ballers to ever come out of SoCal. He played at Ventura Juco with Cedric Ceballos, then went on to Seton Hall before blowing out his knee.

Now he's back on his home turf, hustling for a living. Travel, real estate, entertainment. He's been producing an event called "Chocolate Sundaes" at the Laugh Factory on Sunset every Sunday night for a couple of years now. Hosted by his childhood friend Chris Spencer. Yeah, that Chris Spencer of the talk show *Vibe* who was the best example of how tough it is to host a talk show until Magic Johnson came along and made Chris look like Johnny.

Anyway, years ago in my never-ending search for basketball players for the movies, I'd been given Pookey's number. I'd call him from Charlotte (*Eddie*)—he'd give me a couple names. I'd call him from Seattle (*The Sixth Man*)—he'd give me a couple names. I'd call him from Santa Monica (*White Men Can't Jump*)—ditto. Like I said, Pookey's got it going on.

That's why I'm sitting in his office. I'm trying to revive my last-gasp screenwriting career. And Pookey's gonna help me. (Only he doesn't know it yet.) So are Maya and Rasha. 'Cause they're sharp, these two. They're impressive, and so is Pookey for hiring them.

Maya hits Pookey on his two-way. He's 20 minutes out, finishing up a renegotiation on a TV deal. I'm happy to wait—in the company of these two women; Pookey can take his sweet black-ass time. Rasha and Maya and I hit a nice riffing rhythm between phone calls, fax replies, birthday reminders, and ticket requests.

And these things I learn; Pookey's got an LLC (limited liability company). He's got a lawyer, but does a lot of his own negotiating. He's just finished talent-producing two TV variety shows. He's working on something new with the William Morris Agency. He's a true showbiz entrepreneur with great connections to black entertainers. He's also into real estate—owning houses in South Central and New Jersey and points in between. He's working on an elite, all-inclusive L.A. travel package. He's looking to launch his own comedy club. And he's still the same old Pookey.

Then we hear him on the stairs, shouting up, "Honey, I'm home!" He appears in the doorway. I rise to greet him. He truly is five foot three—wearing a sleeveless denim shirt, baggy jeans, a big smile. He's rough, Pookey. He's not some smooth-polished dude. But I've had enough of them the last few years, black and white. I'm looking for an ally who gets things done.

We shake hands and share the obligatory one-shoulder hug. Then he pulls back and looks me up and down.

"Rob, my man, what've you got?"

"Two things," I say. "Let's sit down."

2

RED AUERBACH STRIKES AGAIN

Pookey leans back, face neutral, as I start my spiel. "First, a screenplay. It's a sex comedy, ensemble—four black actors, three white—a young black woman gets dragged along by her friends on a river rafting trip. Paramount optioned it, but then passed and I got it back and it deserves to get made."

"What do you want me to do?" asks Pookey.

"Look, here's a list of potential actors."

I hand him a couple dozen names—Jamie Foxx, Gabrielle Union, LL Cool J, Jada Pinkett Smith, Marlon Wayans, Beyoncé—like that.

Pookey peruses.

"I personally know more than half the people on this list."

"Cool," I say. "Help me put together a cast and you'll get a producer credit."

Rasha—Pookey's exotic Egyptian-American assistant—sits listening to my pitch, taking notes. Maya, Pookey's exotic Indonesian-American assistant, keeps working the phones in a hushed voice.

"Anyway, here's the script. First you've got to read it and like it. Then if you're interested, we'll do a handshake agreement—you attract some meaningful actors, you're in as a producer."

"I'll read it this weekend," says Pookey, and a little bell in my head goes off—how many times have I heard that?

Rasha speaks up, "You said you have something else?"

"Yeah," I say. "And this one's more immediate. Think the Harlem Globetrotters married to the street feel of the And 1 Mix Tape Tour, throw in *Drumline* and *Bring It On* and it's gold, man. Plus I've got a mystery element that's gonna blow everyone's minds."

Pookey looks puzzled.

"I don't get it. What is it?"

"It's a basketball, music variety show. I'm callin' it *Hoop de Ville*. It's a live show. You stage it in arenas or bring it right into a theater. Plus we might create some reality TV out of it."

Pookey considers.

"Yeah, but, the Globetrotters . . . Rob, man, that's like one of the strongest brand names in the world. And And 1? They created that street cred—it didn't just happen. They worked it."

"Yeah, right."

"So what makes you think anyone's gonna give a shit about another basketball show?"

"Because I'm introducing new elements. Stuff the Globetrotters and And 1 don't have. Look, I started working this idea way back on *Blue Chips*. You saw that, right? Shaq, Nick Nolte, Penny—college hoops."

"Yeah, yeah . . ."

I feel myself suddenly sliding off balance—*Blue Chips, Blue Chips*. Bad memories come swirling back like giant nasty locusts. What a nightmare job. First off the director, Billy Friedkin, was the guy who directed *The Exorcist*, and it was like he hadn't gotten it out of his system. Plus my buddy Ron Shelton (who directed *White Men Can't Jump*) wrote the *Blue Chips* screenplay and was producing, and he and Friedkin didn't see eye to eye and I knew I'd get caught in the middle. (Ron was letting Friedkin direct it because it was the only way to get it made—seeing as how Friedkin was married to Paramount president Sherry Lansing.)

Blue Chips—the job from hell.

Example—in the middle of the movie Nolte suspects that one of his players may have shaved points; so he goes back to the videotape to review the game. Which means it's a game we've got to shoot. Piece of cake, right? I do a couple of casting calls at the Hollywood Y, hire 20 players, a great mix—most of them are black, most of them played college ball, even some D1 in there. Nice size, they're in great shape, they look like college guys, and I figure they're just what's required for the ten seconds of videotape we need for the scene.

I tell Friedkin we're good to go, but no, he wants to see them. Not only that, he wants Red Auerbach and Pete Newell to check them out as well. It turns out that Friedkin's brought on the two octogenarian basketball Hall-of-Famers to guarantee the verisimilitude of the basketball. So we rent a gym, we bring in the players, plus Red and Pete (Red as crusty as Pete is gentlemanly). And I run a 10-minute scrimmage. After which Friedkin turns to Red Auerbach. "So Red, whaddaya think?"

"These guys can't fucking play," says Red. "They stink, the whole bunch of 'em."

Oh, man. From behind, I see Friedkin's neck flush with anger. His head swivels, his eyes lock on me, and I'm thinking here comes the projectile vomit.

"Can you handle this job?" he asks in a steady voice.

"Yes," I answer.

"Then get the fuck out of here and bring me some real ballplayers."

I stagger out of the bleachers thinking, Thanks a lot, Auerbach. This ain't the NBA, man. We're re-creating one mediocre college basketball team here for God's sake. Ten seconds of videotape.

Two weeks later we get the word—Red Auerbach has suffered a heart attack and won't be able to stay with us for the duration of the movie. Was I relieved? Yes. Did I feel guilty about that? No. Because Red, who tormented many a soul during his lifetime, wasn't done yet. He recovered nicely, but thank God slowly. A couple weeks later we shot the scene using the exact same players (Friedkin never knew the difference) and it all played great.

Where was I? Oh, yeah, pitching Pookey *Hoop de Ville*.

I jabber on, "So anyway, in *Blue Chips* we created several college games—pep bands, mascots, cheerleaders; Shaq, Penny, plus 14 first-round draft picks—and I realized, this isn't just sports, this is theater, this is an incredible show."

"Yeah, college basketball," says a skeptical Pookey.

"But this is showball. And we're bringing it to the next level," I say. "Look, the financials are very promising. It's cheap to produce. You can travel it, you can find a permanent home. Like Branson, Missouri—you know, the mecca of country music, it's where . . ."

I see Pookey's eyes begin to glaze over.

"I'm thinking of calling Yakov Smirnoff," I plow on. "You know, the Russian comedian—he's got his own theater in Branson. It's the Midwest, man, they love their hoops out there and—"

Pookey's two-way buzzes and he checks it and I'm thinking, TMI (too much information) . . . Keep it simple, man, and get out the door. Or better yet, lie.

"Look, I'm raising some money," I say (immediately reclaiming Pookey's attention without actually lying). "I want to workshop it here in L.A. I'm gonna need a choreographer, a musical director, an M.C."

"That I can get you."

"That's all I'm asking. So, here's a three-page description of the show. And here's my screenplay."

"I'll read it this weekend," says Pookey.

We say our good-byes and I flush back out onto the Melrose Avenue sidewalk, where the hipster parade marches on—all sorts of sweet young things, showing skin and thong bait—but all I'm thinking is, Man, I gotta refine that pitch. Look, *Hoop de Ville* has great promise. I believe in it as much as in anything I'm doing. I remind myself that sometimes the best ideas are the hardest to convey. Can you imagine the blank stares the creators of *Avenue Q* got the first time they tried to pitch it—"See, there are these kind of Muppet puppets, but they're all real horny, and they like sing and stuff and the actors are onstage with them and . . ."

Monday morning I call Pookey. He doesn't return. I send him an email. No response.

Same thing Tuesday. And Wednesday.

I think, Hey, the guy's a producer, right? Don't take it personally.

Thursday morning I get a call. It's Nigel Miguel. He played at UCLA, got his 15 minutes in the NBA, then bounced around the CBA and Europe before hanging it up. Nigel's from Belize—cool and quiet. He's been supervising Nike commercials for the last several years. Always working on something—shuttling between L.A. and Belize, where he's developing a resort. He was a player in *White Men*, teaming up with Duane Martin in the scene where Woody finally dunked. (Exactly how high was that rim??? Ah, barely over nine feet. But it looked real, right? It looked real.)

"Nigel, man, how are ya?"

"I'm good, man. Listen, I'm sitting here with LeBron James's agent, and you know, they're looking for a project for LeBron, and I thought of that basketball movie you were talking about."

"Yeah, right," I say. "*94 Feet of Hell.*" It's the story of one college basketball game, start to finish, from every conceivable point of view. The movie never leaves the arena.

"Anyway," says Nigel in his deep, quiet voice. "I'm doing a commercial tomorrow with Gary and you oughta come down and meet Aaron."

I'm thinking, Gary, Aaron??? Oh, yeah, Aaron Goodwin, that's LeBron's agent. He and his brother work out of Seattle and Oakland—that means they handle Gary Payton, who's just joined the Lakers.

"Yeah, sure," I say.

"Bring the script," says Nigel. "These guys are serious."

"Sure, Nigel, sounds good. Email me the directions, will ya?"

And I hang up thinking, Pookey? Pookey who?

3

WORKING LEBRON

I pick up the 110 South off the Hollywood Freeway and grind past the towers of downtown L.A. Off at Ninth, west on Olympic, and I see the trucks. It's a PlayStation commercial that Nigel's coordinating, which means they'll be spreading some real money around.

I park my crappy old Toyota Camry around the corner, grab my bag, and walk along the chain-link fence, surveying the scene. It's low-key, a smattering of crew members, ad agency reps—cameras, cables, video monitors under the hot hazy sun. Gary Payton's alone on the sloping asphalt court. Shooting little jumpers, bang bang bang—ball through chain. He's in a rhythm and he doesn't miss. He's got that aura—a pro's pro.

I spot Nigel. I wait until the A.D. yells, "Cut!" and I step through the gate and head right past Payton. I met Gary, years ago, on that turkey called *Eddie*—you know, the one where Whoopi coaches the Knicks. It was supposed to be a comedy, but you wouldn't know it by the too-many-screenwriters script, plus someone forgot to tell the director.

Anyway, we were in Charlotte, North Carolina, pulling together this imaginary Knicks team. It was the NBA summer lockout (1998) and Kurt Rambis and I were handling the hoops. NBA management made it clear to Disney and the producers: While the lockout's on, it's either us or the players. Somehow a movie where Whoopi Goldberg bursts into the locker room and shouts out, "How's it hanging?!" and you cut to a naked David Stern instead of John Salley wasn't exactly what anybody had in mind, so Disney went with the players (although Walt would've rolled in his grave either way).

In retrospect, it's too bad the NBA didn't shut us down entirely, saving the moviegoing public from one more half-baked sports comedy. And if you think I've just got a personal ax to grind, here's what Mr. Cranky (the Internet critic) had to say: "*Eddie* is a good example of the utter bankruptcy of creativity and originality that is Hollywood. This film has all the energy of a rotting corpse." See, I'm not alone.

Rambis was done as a player but not yet hired as a coach, so he was free to work on the movie. Years later, down in Dallas with the Mavericks, Del Harris said to me, "You know, I was supposed to do that movie but because of the lockout, Rambis replaced me, then he did it again for real (as Lakers head coach) the very next year."

Del seemed pissed about it, but the truth is, neither of those guys could run the psyche game with Shaquille O'Neal.

Kurt had a helluva Rolodex, though, and this imaginary Knicks team was coming together—John Salley, Rick Fox, Malik Sealy (R.I.P.), Dwayne Schintzius (remember him?), and Greg Ostertag—but we still needed a point guard. The producer, Mark Burg (whom Ron Shelton once slammed up against a wall during the shooting of *Bull Durham*), wanted Gary Payton in the worst way. I had my doubts—movie shoots are long and tedious—and Payton didn't suffer fools gladly (which would've made this movie particularly tough on him).

We were working out in a practice gym—Salley, Fox, Ostertag, Sealy, Schintzius, and I was thinking, Spike Lee is gonna bug when he sees these guys portraying his beloved Knicks—but given who the Knicks had been putting on the floor lately, our guys just might've eaten their lunch. Anyway, Rambis and I were choreographing some plays—it was all loosey-goosey, lots of laughter, when suddenly the double doors swung open. It was Gary Payton, and that fast, the vibe changed.

Payton's a thoroughbred—high-strung, tightly wound, quick to bristle. He walked in, said a few quiet hellos, went into a brief conference with the producers, took one more look at the players on the floor, and walked out the door. Good decision. That afternoon Kurt called Mark Jackson and we had ourselves a point guard.

Back on the commercial set, I slide past Payton (no sense trying to say hello right now—he wouldn't remember, he wouldn't care) and head for Nigel. Nigel's six foot six—slim and dark-skinned, his head shaved. He always wears shades and always plays things cool and quiet. We shake hands, bump shoulders.

"Rob," he says, "this is Eric Goodwin."

I shake hands with the handsome, well-built black man. He's dressed sharp, in baggy designer pants and a knit top. His head is shaved, his manner direct. His brother Aaron is his identical twin and they've pulled off the coup of the year—they represent LeBron James.

"How you doin'?" asks Eric.

"Not as good as you," I reply.

Nigel says, "Rob's got that project I was telling you about. *94 Feet of Hell*. About that one college game?"

"Oh, yeah. Yeah," says Eric.

"If you guys are looking to do some movie production, get LeBron involved, this is a great place to start," I say, thinking I'd better blurt it out fast since I don't know how much time he'll give me. "For one, it's pure hoops—a hard-core, inside look at the game; for two, it's a quick shoot, under three weeks next summer, so you wouldn't be getting tied up for too long. Plus it's basketball, man—a great place for LeBron to work up his acting chops. It's what he knows, it's where he's comfortable." (I almost say it's also a great place for him to get a taste of college ball since he's straight out of high school, but I leave that out.) "Plus we'll put together some of the best young players in the world, so he'll be able to work out every day."

"Just like in *Blue Chips*," says Nigel.

"Yeah," I add. "We had 14 first round draft picks in that movie."

Eric Goodwin smiles and says, "The *Blue Chips* curse."

"How's that?" I ask.

"Thomas Hill. Rex Walters. Adonis Jordan. Ed Stokes. Eric Riley. They all disappeared."

"Not to mention Bobby Hurley," says Nigel, and I'm thinking, Oh, man, ease up, don't turn this into a train wreck by association here.

"Somehow," I say, "I don't think that's gonna be LeBron's problem."

Eric Goodwin smiles. When you've got a client like LeBron James, you want to navigate his career course very carefully. "It sounds good," says Eric. "But you know, we're just feeling our way right now."

I deflate a bit.

"Rob's tight with Ron Shelton," says Nigel.

"That's good," says Eric. "Ron Shelton, he's got that touch, you know what I mean? That's the kind of director we'd want for LeBron."

And I'm thinking, Yeah, yeah, here it comes, and sure enough, here it comes.

"Any chance you could put us together with Ron?" asks Eric. "You know, LeBron, me, my brother. Just to talk."

"Sure," I say. "Let me run it by him." I'm thinking Ron will be open to this. Sitting down in his office one morning with the most acclaimed player ever to come straight out of high school.

"Good," says Eric.

The crew is breaking down the set—moving to a new location.

Eric, Nigel, and I start walking to a trailer, and the clock is ticking. "Look," I say. "You guys are gonna be entertaining a lot of movie possibilities. But this one's clean. It's direct. And it's not some lame comedy. Plus with LeBron in it and you guys involved as producers, we can go straight to Turner or ESPN, and you know they're gonna jump at it."

"You got the script?" asks Nigel.

I pull it out of my bag. "By the way," I ask Eric, "do you know Pookey Wigington?"

"Pookey?" says Eric. "Sure, I just talked to Pookey yesterday. Pookey's the man. Why?"

"Oh, Pookey and I are working on something together."

"Cool," says Eric.

I hand him the script.

"I'll read it this weekend," he says.

4

RAYS OF HOPE

Eric Goodwin says good-bye and steps inside the plush trailer to join his identical twin brother, Aaron. The two are da bomb (or is it da bombs?)—polished, intelligent, ambitious young African-Americans, taking care of business. We're not talking Tank Black here. This ain't no Master P. I turn to Nigel.

"That went nice," says Nigel.

"You think?"

"Come over to our next location. Ray Allen."

"Cool," I say.

Nigel hands me a map. The gym's over at a Methodist Church on Wilshire. I drive over, park on the street, and wander down alongside the church until I spot a couple of grips carrying equipment. The gym's upstairs. I walk in and climb the cold metal stairs. The gym is old and cramped and worn, but the backboards and rims are up, and there's white sunlight spilling through some tall dusty windows.

This gym is a church. It's not in a church, it is a church. Then again, every gym is a church. It gets to you when you let it. When you're alone with it. If you love basketball. If you've put in some time with the game.

When you walk into an empty gym—it doesn't matter if it's a junior high school or Madison Square Garden—there comes this wonderful sense of expectation, that all things are possible, mixed with a twinge of anxiety, followed by a flush of well-being. Or it can be just a hoop even. Outside. Driving along in the country—some farm kid's nailed a rim to a telephone pole. Or the side of a barn. At the end of an alley. In the parking lot of a mattress store.

If you look for them, you see these signs—these implements of worship—everywhere, and you see the followers of the faith, enraptured with sweat and effort, worshipping in the Church of Hoops. You don't have to believe in God. You have to believe in Michael.

Blah, blah, blah. The crew's dragging in cameras, pulling cable, mounting screens to further diffuse the light. The second A.D. appears and yells, "That's lunch, half an hour."

I catch Nigel on the stairs, talking on his cell. He motions and we walk out and head across the parking lot to the lunch wagon. Crew eats first. Always. We linger, then join the end of the line. How's this sound: Grilled halibut, topped with fresh ceviche (scallops, shrimp, and rockfish in a spicy salsa). Rib-eye steaks, grilled to order. Rice pilaf. Fettuccine alfredo. Potatoes au gratin. Carrots and peas. Asparagus spears slathered with hollandaise sauce. An endless salad bar. Breads, rolls, muffins. Cakes, pies, ice cream, coffee, soda.

And to top it off, as we're finishing this typical lunch, someone comes around with warm cookies and chocolate milk shakes in huge plastic cups.

Ray Allen arrives. Nigel and I say hello and shake hands. Ray gets some food and sits down the way a bit with some of the crew. Everyone's cool. No one's rushing him.

"Too much fat in this meat," he says, pushing the plate away and concentrating on his vegetables. The man is in perfect condition. His body fat must be like point -3 or something.

The second A.D. strides by. "We're back. We're in."

The crew slowly stand, dump their trays, and head back into church. A pretty, nearly middle-aged woman from wardrobe looks over from the next table.

"You must be Ray," she says.

"That's right."

"I'm gonna get you dressed," she says.

"You make it sound easy," says Ray.

They stare at each other and smile. A sweet exchange, not even flirtatious really. Just Ray being Ray.

Up in the gym, they're still setting cameras for the first shot. Nigel's standing in for Ray at the foul line. Ray's sitting in a director's chair along the baseline near the video monitors. I wander up.

"Hey, Ray, have you done any movies since *He Got Game?*"

"Just *Harvard Man*," he says.

"I somehow missed that one," I say. He doesn't smile. "Have you seen *Playmakers?*" I ask.

"It's good. Man, that show is good."

"I think so too," I say.

"It's like they get it," adds Ray. "The dynamics . . . all those shades of gray."

"They really nailed the casting," I say. "I buy those guys. My only quibble, you watch that show you think every player in the league is all messed up."

"What do you wanna watch, good guys?" says Ray. "Good guys are bad TV."

Later, I slide over to Nigel. "Hey, Nigel, how long's Ray been in the league?"

"Uh, seven, eight years . . ."

"Think he could play a college senior?"

Nigel looks over at Ray. "With that face? Easy."

"*94 Feet of Hell*," I say. "There's a great part for him. A pure shooter who missed a big shot."

"I already mentioned it," says Nigel. "He's gonna read the script this weekend."

In Hollywood, nobody reads. All these waitresses and valet parkers running around slipping people scripts, it's ridiculous. 'Cause nobody reads. They don't even read if that's their job. Agents don't read, actors don't read, studio executives don't read. Get it through your head: Nobody's gonna read your script. They might buy it, though.

Nigel gets off his cell phone. "Damn," he says. "Nike needs me in Sacramento tomorrow. We're shootin' LeBron."

"So?"

"So, could you cover for me? Up in Burbank. It's a spot for the Professional Bowlers Association. Six guys playin' hoops with a bowling ball . . ."

"And for this, I would be paid how much money?" I ask.

"Five hundred straight up," he says. "Plus they said if they worked me into the scene, another five."

"I'm in," I say.

"They're gonna blow up a backboard," adds Nigel.

"Stunt adjustment," I say. "Hazardous duty."

I hate explosions. I hate pyrotechnics. Too unpredictable. But if God didn't want people to use explosives, he wouldn't have created the Chinese. Way back when, I worked as a locations scout on *The Warriors*. We needed a street to blow up a Cadillac. I found one out in Brooklyn someplace, but Walter Hill, the director, said it looked too open, so they sent me to some trucking joint over in Queens where I rented six huge tractor-trailers and had them driven over and parked in the street.

Then Walter didn't like how bare they looked so they sent out three union scenic artists who spent an entire day painting some lame-ass graffiti on one half of one side of one truck. Talk about milking a job. We all went home that night, the next morning we're back and all six trailers are slathered in bold graffiti, front to rear, top to bottom.

That night they wire the Caddy to blow it up. It's tense on the set. We were striding up and down the street, making sure it's all locked down, the neighbors out of harm's way. Way down the street there's this old Italian couple, must have been in their 90's who wanted to watch. They'd set up lawn chairs on the sidewalk. I asked an A.D. if we were safe there and he looked at me funny.

"You're like 300 yards away."

"Yeah?"

"Of course you're safe," he sniffed.

They finally rolled cameras, yelled "Fire in the hole!" (they actually yell it), and the car blew up. The huge steel hood came screaming straight at us on a line drive, flew right over our heads, and smashed into the building behind us.

"*Buono*," said the old signora.

But the old man, he was pissed.

5

BALLING WITH BOWLING BALLS

So I show up to coordinate this commercial. An easy 500 bucks. A piece of cake. It's some streetballers going three-on-three in a park. But with a bowling ball; no dribbling here. They're supposed to run a little play, though—pass and screen away for a teammate, who curls into the paint, catches the bowling ball, then flings it up and through a glass backboard, which explodes into smithereens.

I haven't read the script, but the gist of it is: Look, bowlers are athletes too—or some crap like that. Not that it matters. The whole issue of what constitutes an athlete only exists because sports editors need to fill space. It can occasionally make for some lively conversation, though.

Here's one I'd never heard. I was hanging out on the set with Ron Shelton. It went something like this—someone said, "Michael Jordan's the greatest athlete in the world."

Ron immediately fired up, "That's bullshit!!!" he growled. (For Shelton, it's not worth discussing if you can't growl about it.) "Michael Jordan is an amazing physical specimen who advanced one game to its highest level. But no way he's the best athlete in the world."

"Who is then?" someone asked.

"Why not Danny Ainge?"

"Danny Ainge? Danny Ainge!!???"

"You mean Danny Ainge the whiner?"

"That Danny Ainge?"

"Yeah," said Ron. "For starters, Danny Ainge has two NBA rings as a member of the Celtics. He shot almost 50 percent from the floor."

"What'd Jordan shoot?"

"About the same," said Ron.

"So you're sayin' Ainge was as good as Michael?" someone foolishly inquired.

"Did I say that!? I didn't say that!" yelled Ron.

"Bob Cousy only shot 37%," I threw in to deflect the heat.

"And they say players today can't shoot," someone added.

"Look," said Ron. "We're talking greatest athlete, not basketball player. Athlete. And before the Celtics, Danny Ainge actually played Major League Baseball. Three years with the Blue Jays. Something Jordan didn't come close to."

"Yeah, but Ainge couldn't hit either."

"It's a matter of degree," barked Ron. "Ainge was twice the baseball player that Michael Jordan was. And last—he's a scratch golfer. Do you know how tough that is?"

I didn't know (having sworn off golf years earlier as just one more unnecessary pursuit that was bound to make me crazy) but I knew we were gonna find out. Shelton, you'll remember, directed *Tin Cup* and is a fine golfer himself.

"How tough?" someone asked.

"Ask Michael Jordan," answered Ron.

The PBA shoot is at a little pocket park in Burbank, just up from the Disney lot. Nigel's strolling across the court, talking into his cell phone (is he never?). The director and crew are off on a tennis court, shooting a sequence where a player tries to serve a bowling ball. It's really stupid, but it's a commercial and you know before you hit the remote you're gonna wait and see what happens so in that sense it's gonna work better than 95 percent of the commercials out there.

Nigel nods hello and pulls out a couple of pages of storyboards for the basketball sequence. They're crude shot-by-shot drawings of how the scene will unfold. Shot—Player #1 with bowling ball at top of key, being defended. Shot—Player #1 passing ball. Shot—Player #2 catching the ball and falling on his ass. Like that.

On the back, Nigel's scribbled out a plausible play that gets all the players to their spots for the last shot, where Player #3 throws the bowling ball through the glass backboard. (And the last time you saw glass backboards in a city park was when??? Not that it matters. It's a commercial.)

The players start to straggle up.

"Where'd you find these guys?" I ask Nigel.

"Two of them are actors. They got cast. The other four, you know . . . here and there."

Two young white men come through the gate, one about five foot seven, five days' growth, Italian-looking, the other tall and fair like Keith Van Horn but even softer. Then four black men—tattooed, muscled—show up with that slow walk onto the court that says they're at home.

"Lemme guess who's who," I say to Nigel, and we both laugh.

Nigel introduces me and the short white guy immediately pulls me off to the side. "Look, I'm not really a basketball player but . . ."

"It's cool," I say. "It's a commercial. We're playing with a bowling ball."

"Yeah, yeah, okay," he says. "I did play some football."

They all played some football.

Nigel checks his cell for the time. "I gotta go catch a plane."

I pull the players together. "All right, let's take a look at this."

We rough out the scene, and right away I can tell that one of the authentic black players (the biggest, meanest, guy of course) doesn't like taking direction. Oh, man. Why do people take jobs like this if they don't like being told what to do? Maybe it's a coach he once had. Maybe he's just hungover. Maybe he doesn't like white people. "It's like this," says my friend, K.B., or "K to the B" as we call him. "It's like my cousin back in North Carolina. He just doesn't like white people. Just like that. He don't like y'all. And with him, if you're white, that's where you gotta start."

So maybe that's it. Or more likely it's that NBA syndrome. Every good basketball player I've ever met is certain he belongs in the NBA. Instead, when you find yourself in some crappy park in Burbank in the hot sun working as an extra in a commercial, it can make you a little surly. This guy is surly.

Maybe I'll tell him Danny Ainge is a better athlete than Michael Jordan.

But suddenly I've got a bigger problem—it's some scruffy-looking guy walking up from the tennis courts.

"Where's Nigel?" he asks.

Uh-oh. Luckily, I'd had a moment to watch the tennis sequence and figure out that this was the director. You can be on some sets for an hour before figuring out who's directing the thing. The assistant director is usually the guy barking all the orders, then there are always all sorts of people hanging out acting important—producers, associate producers, coproducers, executive producers, line producers—and most of them are dressed like they're still in college so it's hard to tell who's who.

Some guys you know right away—like Joe Pitka, who's directed some of the great Nike commercials. I've never had the pleasure but I hear he can be a real asshole. A screamer. And probably proud of it. I'll have to ask Nigel sometime, although Nigel's too politic to go around telling stories out of school.

Anyway, this young scruffy-looking director comes walking up asking for Nigel and I realize the producer hasn't told him I'm replacing Nigel for the day. Is he a screamer? I'm about to find out.

6

THE FLYING STUNTMAN

"Where the hell's Nigel?"

"Nigel had to leave," I say, looking him in the eye and quickly adding, "I did the basketball on *White Men Can't Jump*, *Celtic Pride*, *Blue Chips*, and a bunch more. And this is gonna work great. No problems."

He stares at me. Considering. People who blow up at unexpected news usually have the luxury. A no-cut contract. A big budget. Powerful boosters. But for most people in charge, you can waste a lot of time and energy throwing a temper tantrum or you can keep things clipping along and get the job done. On a one-day shoot like this, with a couple dozen camera setups to grind through and a sun that's going to fade fast come five o'clock, this young director makes the wise choice.

"Let's see it," he says.

"Okay, gentlemen, please go to number one," I say loudly.

The players all hustle into position except for the one surly one, but the director doesn't notice because I've buried him in the paint, setting screens and taking up space. That's what you get for having attitude. You could've been a star, man. You could have been the big mean dude throwing that bowling ball through the glass backboard if you'd only been a little more cooperative.

"We'll do it at half speed first. Then run it again, full speed. Ready? And action."

The players run through the play with a basketball subbing for the bowling ball. Then again, and the director sees that it's gonna work fine.

"There's one thing," I say. "It might be smart if the shooter can work over on the grass with a real bowling ball. So he can find his range."

"Nah, I want it natural," says the director, and he walks away.

I head over to the special effects crew, who are rigging the far hoop with a special glass backboard.

"How many of these things do you have?" I ask.

"Three," a grip answers.

"What's the changeover?"

"About an hour once we blow the first one. We can't set the explosives until it's up in the air."

Explosives. Here we go again. I remember another incident back in New York when I was scrambling in movie production. I'd gotten a call to help out on a Sean Connery movie, *The Next Man*, which ranks right up there on that "Crappy Movies That Never Should've Been Made" list. Connery was playing an Arab diplomat and they'd planned this big scene where there was an Arab riot in front of an embassy on the Upper East Side and they were gonna blow up a limousine.

The production manager asked me to help with crowd control for the day and when I showed up on East 92nd Street they put me in a leather jacket and a Yassir Arafat burnoose and stuck me right in the middle of all these half-crazed extras they found somewhere out in Far Rockaway or someplace and we stood there for hour after hour waiting for them to rig the limo and everyone was getting pissy 'cause it was freezing out and they wouldn't even let you leave to take a leak. The A.D. rushed over (A.D.s rush everywhere—it's the only speed they know) and said, "Listen, we're losin' the light. We're not gonna have time to rehearse this, so when we yell 'Action' make sure it looks real, right, no laughing, no fucking around, all right?"

"What's taking so long?" an extra asked.

The A.D. ignored him and rushed off. "Fuck you too, pal!" the extra shouted after him.

What was taking so long was it wasn't enough to just blow up the limo, these guys wanted it surrounded by three stuntmen rigged with special harnesses so that when the bomb went off, steel cables would yank them up and away from the explosion like they'd been blown up. Now this is hairy shit under the best of circumstances, but on a cold November day on East 92nd Street with a mob of pissed-off extras and the light fading fast, it was a recipe for disaster.

I hate explosions, I hate special effects—have I mentioned that?

Anyway, the three stuntmen were really scrambling. Making sure the steel cables they were wired to were running away from the limo at the right trajectory, that the explosive canisters that would actually trigger the cable-yank were good to go, and that all three were synchronized to go off with the limo bomb.

One of the stuntmen was stretching his cable right through the mob I was part of.

"You're with us, right?" he asked. "Production?"

"Yeah," I said. "So where you guys gonna land?"

The stuntman looked from the limo to the far curb.

"Uh, somewhere between here and that building."

The A.D. suddenly screamed out through his megaphone, "If we don't get this shot in the next three minutes, we're dead! You hear me, people? Dead!"

The stuntman looked over the situation again.

"Listen," he said to me. "When that bomb goes off, I'm gonna come flying right through here, all right? Like right over your head, okay?"

"Yeah . . . ," I said.

"So reach up and try to knock me down," he said. "So I don't hit that curb back there."

"Okay," I said.

The stuntman rushed off. The whole scene felt hyperwired—people hustling everywhere, jostling, elbowing, snarling, screaming instructions. Finally, the A.D. called, "Roll cameras!" and about six different Panaflexes cranked up, camera assistants holding up clapboards and shouting the scene number before ducking away. Then the A.D. called for background action and all the extras started shouting and surging forward and suddenly BOOM!!! the limo blew up and there was smoke and sawdust and cork pieces shooting everywhere and that fast, there he was, the stuntman, hurtling through the air right toward me at about 9,000 miles an hour.

I hit the deck.

He screamed past right overhead.

The whole street went silent. The explosion had knocked the air right out. A second plume of smoke rolled off the limo and through the crowd of fallen extras. It was surreal.

A voice yelled out. "Cut! That's a cut!!!" You could hear a murmur/mutter spread among the crew and extras: "Holy shit."

The A.D. shouted through his megaphone, "Okay. That's a wrap! That's a wrap, everyone! Go to wardrobe! Sign out and go home. Thank you very much! Thank you! And don't steal anything on the way out!"

I looked through the mayhem to the far curb. The stuntman was crumpled up in a ball. He slowly unfolded his body, feeling for damage. His face was black with gunpowder and burnt cork. He rubbed his eyes. I walked over and he looked up at me.

"What happened to you?" he asked.

"I ducked," I said.

"Smart move."

Back on the bowling ball commercial, we bounce through a bunch of camera setups, then it's time to blow the backboard. The A.D. calls a safety meeting. The special effects guy explains there might be tiny shards of glass so we should all protect our eyes. One of the players says he was looking at the hoop when the shot went up, so what should he do now. It's a good question.

The director and cameraman consider. "Look," says the director. "We can cheat you back a bit. You won't be right under it."

He's not exactly answering the question. As the cameras are set, I pull the players together. I'd just learned that this isn't a union shoot. There's no SAG (Screen Actors Guild) minimums here. No overtime. No stunt adjustments. No doctor on the set.

"Gentlemen, listen up," I say. "Just before the ball hits the backboard, turn your heads away."

"But . . . "

"It's a commercial," I say. "You're willing to lose your eyesight so more people will watch bowling on television?"

So they roll cameras, we run the play, the bowling ball smashes through the backboard (a little high and wide but at least he doesn't miss), the explosives go off, the backboard shatters, and nobody gets hurt.

At wrap I pull the producer aside.

"We need to talk money," I say.

"What did you arrange with Nigel?" he asks.

"He told me a thousand bucks. He said you told him 500 for the job and another five for using him in the scene."

"We didn't use you in the scene."

"I wouldn't have been right," I say.

He stares at me.

"Look, call it 750 and we both walk away. You're happy with the job, right?"

"Yeah," he said.

"So you just saved 250 dollars."

Nigel hits me on the cell as I'm driving home. He's up north, working that Nike spot with LeBron.

"Tell me it went fine," he says.

"It went fine," I say. "How's Sacramento?"

"It's Sacramento. Listen, man, about this column you're writing. I'm a little concerned. I got these projects developing with the Goodwins, you know, and that means yours too, *94 Feet*, and I don't want to jeopardize anything."

"It's cool, Nigel," I tell him. "Everything's gonna be cool."

But I hang up wondering, What am I doing here? It's one thing to chronicle your business efforts in a book that will show up a couple years later; it's another to be running a column on ESPN.com that's being read by tens of thousands of people every week. But the column is a gas. I've had four posted already. And at 500 bucks a pop, that ain't bad. But every week, I expect that they're gonna pull the plug. That some Disney honcho's gonna read it and say, "What the fuck is this?" So far, so good. I'm planning on keeping my head down and keep churning them out. It may not have been so wise to post my email address, though. The first response I got? "Dude, not to be rude, but what the fuck was that?"

7

CAN HOLLYWOOD HANDLE THE TRUTH?

Oh, man, what's going on? It's been ten long days since I sent an email to LeBron's agents at Goodwin Sports Management, and not a word.

Each morning I open Outlook Express with a mixture of dread and hope (in that order). It's not the email grenades that worry me—the ones that say, "Got to be the dumbest story I have ever read, I want my ten minutes back."

No, it's my concern that I'm pissing off the very people I want to be working with.

Like Mark Burg, one of the producers on *Eddie*, which I recently called a piece of crap. Is he going to hold it against me? We'll find out. I hope not. Making a good movie is one of the hardest things you can do. People fail. A lot of honest attempts go awry, and you can't have too thin a skin if you're going to make a career of it.

Plus Burg actually has a pretty good sense of humor. Like the time we were prepping *Eddie*. We hadn't left for Charlotte yet; we were working in one of Disney's sterile high-rises in Burbank. Whoopi was visiting and she was one very unhappy camper.

You learn to keep your head down in the movie business. Doesn't matter what department you're in, as soon as the shit starts flying you close your door (if you're lucky enough to have one). So there I was, alone at my desk, trying to figure out how in hell, in the midst of a nasty NBA lockout, we were gonna put actual NBA players in uniforms that belonged to management.

Suddenly I heard Whoopi go off. This wasn't one of those slow builds; oh no, this was an explosion—a monumental seven-expletive tirade laced with what sounded like a string of extremely imaginative quasi-sexual improbabilities about to be committed on said producer, Mark Burg.

I closed my door.

The thing about movie stars—when they yell at you, you can't yell back. You gotta eat it. You gotta bend over and take it like the hotshot producer you are.

Then you turn around and dump it on your assistant.

Whoopi was finally finished. She stalked out. A long silence. I saw my door slowly swing open. It was Burg. His face was crimson. Maybe he couldn't find his assistant (although I never once heard him yell at Darcene, who was black and proud and unlikely to take it). With me he decided to play it for laughs. He smiled wanly and hobbled toward my desk, knees bent, legs spread, gingerly holding his arched buttocks in mock pain.

"You got a first aid kit in here?"

So maybe Burg will be cool. There are a couple of projects we could do together. He's got a hit TV show, *Two and a Half Men*, plus he's got a new movie out, *Love Don't Cost a Thing*, which is worth the eight bucks and this heartfelt plug is a lame attempt to mollify him.

But still, it's tricky territory. So I email Mark Cuban (the billionaire owner of the Dallas Mavericks), expressing my concern.

"What do you care what people think?" he replies.

Easy for him to say, given that he's worth about 16 billion gazillion dollars. "F.U. money" it's called (for those of you under 18, that stands for "Forget You"). Whereas, I'm still driving a Toyota with over 176,000 miles on it.

But Cuban's got a point. And—love him or hate him—he puts it on the line every day. Like this feud he's got going with the Lakers.

Cuban knows that for a rivalry to truly develop, things have to get personal. It's good for business. But then some wacko comes caterwauling out of the stands, and you're suddenly wishing you were back at that Dairy Queen.

Anyway, I decide to buck up, lay it on the line. Figure people like seeing their names in print, even in less than flattering circumstances.

But I just about crapped the day my third column came up and I saw that the knuckleheads at ESPN (you guys know I still love you, right?) gave me a splashy Page 2 headline, "THE LEBRON PROJECT."

Like they were doing me some kind of favor.

Oh, man. There is no "LeBron Project." There is a project I had pitched to LeBron. Actually to his agents, the Goodwin brothers. Actually just to Eric Goodwin; Aaron was in the motor home.

But there was not (as yet) a "LeBron Project." People are very protective of their turf. What was Eric going to think seeing a headline like this?

I fired off an email explaining that yes, I'd taken the liberty of describing the meeting—the one that had been arranged by our mutual friend Nigel, who was also concerned for my relationships in the business. I expressed my respect. Explained my powerlessness over the headline writers—"More Dirt from Hollywood"; jeez, fellas. Said I hoped I hadn't jeopardized the Goodwins' and LeBron's potential involvement in my college basketball movie, *94 Feet of Hell*.

Days pass. No reply.

I call my friend and fellow screenwriter, Oliver Butcher, for his take. Ollie's a Brit and the closest he gets to sports is bagging ground squirrels with his crossbow, but I value his opinion on all things Hollywood. He's recently returned from Vancouver, where he was doing rewrites on Dwayne "The Rock" Johnson's latest flick.

That night, sitting around with our wives, Ollie tells a titillating tale about life on location. (See, one of perks of being middle-aged guys with thinning hair who don't cheat is that when we tell racy stories in front of our wives, they actually laugh and don't get offended because they know how idiotic men can be and they're just glad theirs got it out of their systems before they got married.)

Anyway, Ollie was up in Vancouver, writing feverishly all day, living the pariah's existence during the off hours. This is partially self-imposed. The most dangerous thing a rewrite guy can do on location is hang out with the actors, because they're always pushing to expand their parts.

"So there I was," says Ollie in his clipped Cambridge accent, "drinking alone at the hotel bar, and I look over and there's this absolutely gorgeous woman staring at me."

"You were drunk," says Ollie's British wife, Sophie.

"I was not drunk. I was living a very spartan existence. I was having a beer. Then she walks over and sits next to me, and she's even more gorgeous up close."

"She had big boobs," says Sophie.

"Hhmmph, she was quite well-endowed, actually," says Ollie. "So we exchange names, pleasantries, and I'm feeling rather puffed up—hey, dude, still got some of the old stuff, say what?!"

"And the next thing you know, she's got her hand on my leg and she's asking, 'Do you have a room here?'

"I say, 'Uh, yes, as a matter of fact, I do.'

"She says, 'For 250 dollars I'll follow you there.'"

We all laugh, and Ollie says, "I was crushed, absolutely crushed."

"What happened next?" I ask.

"She saw the look on my face and she added, 'That's Canadian,' and I bolted from the room."

Later, I pull Oliver aside and explain my dilemma.

"Don't worry, Raab"—which I guess is English for "Rob"—"look at William Goldman. *Adventures in the Screen Trade*. Nobody knows anything. He slammed everybody in that book, and he works all the time."

"Yeah, but that was after he won the Oscar for *Butch Cassidy and the Sundance Kid*."

"Then what about Cecil B. DeMille?" says Ollie. "What did he say? 'I will never ever ever ever ever ever work with that man again . . . unless I need him.'"

So back to Goodwin Sports Management and how I might have blown a relationship. I'd been thinking, They're busy, these guys. Maybe they never even got the email. Hey, Aaron just had a run-in with the Oakland Police Department. Maybe that's it. The DWB (Driving While Black) was demanding all their time and attention. An incident like that is humiliating and infuriating and can really throw a man.

It brings me back to a day on *White Men Can't Jump* when I got my own DWW (Driving While White). Actually I wasn't even driving a car. I was riding a bicycle, which makes it more of a PWW (Pedaling While White).

The memory comes back, vivid as yesterday, and even after all these years it stings like a slap.

8

WHITE MEN CAN'T PEDAL

It was early morning on the Venice Beach courts. We were socked in, waiting for the morning fog to burn off. We'd been shooting only two or three days, and the cast and crew were still finding their footing, working out the dynamics, but this movie was a winner. Even that early in the shoot, we all knew it, and it's one of the best feelings in the world. From day one, Wesley and Woody had the sparks flying.

We had a couple dozen players out there, shooting hoops in the chilly fog, calling out, laughing, scarfing down breakfast burritos and hot coffee. The steel rims clanging, chain nets snapping. The surf a dull roar from across the empty sand. Waiting for the sun.

One of the balls was flat. I looked around but couldn't find the prop guys, so I started walking down the Venice Beach walk toward the prop truck and came upon Kirk, the incredibly crusty sound mixer (aren't they all?) who'd rolled his sound cart as far away from the set as it was humanly possible to and still be working on the same movie.

Kirk had a little BMX bike that he kept to hustle in on when he was needed. I asked if I could borrow it, and since he hadn't had the opportunity to hold something against me yet, he said yes.

I jumped on the bike and sprinted down the empty walk through the fog, and then suddenly I heard a loud voice yell, "Hey, you! Stop!"

I kept riding. Venice is filled with lunatics. When some stranger yells for you to stop, you don't.

Suddenly I heard the *whoop whoop* of a police siren. I immediately got that queasy feeling in my gut. I skidded to a stop.

Two cops drove their black-and-white up the walk through the fog, stopped, and climbed out. One white, the other Latino. The white guy had a hair up his butt.

"Hey, asshole, why didn't you stop?"

"I didn't know who it was."

"You didn't know who it was?"

"That's right."

"I'm a police officer."

"I didn't realize that."

"Do you realize it now?"

"Oh, yeah."

"Oh, what?"

"Yes."

"Yes, what?"

"Yes, I realize that."

"No. You mean 'Yes, sir.' Right?"

"Uh . . . right. Yes, sir."

"Did you realize you were breaking the law?"

"I didn't realize that."

"You don't realize too good, do you? What are you, some kind of dumb-ass?"

I said nothing.

"Answer me, dumb-ass."

And on and on it went. My mouth clenched against the urge to say something really stupid that might cost me one of the coolest jobs I'd ever landed. So I stood there and took it as he systematically chipped away at my dignity until finally he'd had enough and let me go.

I walked the bike back to the set, burning with anger and resentment. It showed. A few of the players came over. Half these guys I knew. We hired right off the courts we played on—Venice Beach, the Hollywood Y.

"Rob, man, what's up?"

"I, uh . . . I just got hassled by a cop."

"For what?"

"Riding a stupid fucking bicycle on the walk."

"Wow, man, you look seriously pissed."

More players wandered up. A couple of them, Reggie and Mahcoe, peppered me with questions.

"They cuff you?"

"Uh, no."

"They make you spread on the ground?"

"No."

"They put a foot in your back?"

"No."

By now the black players were exchanging looks.

"They put a gun on you?"

"No."

There it was, the first smirk.

"They make you go to your knees and crawl backwards towards them?"

"Uh, no . . . "

"Then what'd they do, man?"

"They, uh . . . The guy was an asshole."

"You mean he talked mean to you?"

Then the smiles started busting out, and I realized, Man, I'm cooked—pity the poor little white boy. I tried to save it by saying, "Yeah, he was a real meanie."

But by now these guys were laughing, and then they were doubled over laughing, then slapping fives laughing, and then they were absolutely delirious with laughter and I just stood there like an idiot, half-smiling, and finally when the roars subsided I said, "Fuck all uh yuhs."

And they started laughing again.

Now here's the part where I'm supposed to say, "Look, we know that most cops are doing a good job, etc., etc.," but you know what—today, right now, I'm not gonna say it.

Cops. Cops and coaches. The good, the bad, and the ugly. When it's a coach, that's your choice. That's your choice to be there or not. But when it's a cop . . . Ask Wesley Snipes, ask Virginia state judge Alotha C. Willis, ask Aaron Goodwin.

The familiar burp of a fresh email pulls me out of the memory. I reach for the mouse. It's Eric Goodwin. He and his brother like my college basketball script, *94 Feet of Hell*. No promises, but they want to meet, with or without Ron Shelton.

This is good news.

Eric writes that they haven't shown it to LeBron yet. They want to keep him focused, but they'll get there when the time's right. He ends with a well-deserved boast, "LeBron's been too busy playing better ball than most NBA veterans."

Ain't it the truth? I personally worried about that moon-shot jumper of his, but he's nailing that now too.

I immediately reply to Eric. Glad you like it. Looking forward to the meeting. (Keeping it cool, professional.)

Okay, this is it, now. This is business.

I call Ron Shelton. Explain that I need advice. The plan is to pull together the hottest young players in the NBA to be in the movie. But it's gotta be a quick shoot. These guys get booked up. Camps, commercials, families.

How many shooting days will we really need to make this thing? It's the story of one college conference championship game, told from both sides' point of view. It's a war movie. *The Battle of Algiers*. *Tora! Tora! Tora!* Plus it's the first sports movie where you can't predict the obvious winner.

It's simple logistically because the movie never leaves the arena. Can we light the crap out of it and shoot the whole thing in two weeks? What about multiple cameras? Second-unit directors working opposite ends. Ron says he'll be glad to talk it through. We set a meeting.

Then I call David Lester. Lester's a line producer and he knows this stuff backwards and forwards. He's also the kind of guy you want on your team when things go south (and there's NEVER a movie where things don't go south). We talk about shooting in high-def video. The need for the right arena. The difficulties of working with pro athletes who might treat this less than seriously (God knows, we've both been there). He'll be glad to talk it through.

Then I email Nigel Miguel. I expect him to be key in chasing down the NBA guys we're after. The plan is to first put together the players, whose star quality will then attract some name actors to play the coaches.

It's called packaging. There's no one set way of putting the key elements of a movie together except to say that it usually involves a great deal of begging, arm-twisting, misrepresenting the truth, calling on old favors, outright lying, threatening, more begging, more lying, beating down doors, beating up assistants, and even more begging (not necessarily in that order).

No more time for messing around. Stay focused. Stay hungry. There's a lot of work to be done. Enough with the cheap jokes.

I hear the burp of another email and click on it. It's Ollie, my Brit screenwriter friend. Uh-oh. I took some license relating a story of his in last week's column.

His email reads: "Nice work, Rob. License forgiven. Next time why not use the anecdote about my 22-inch penis."

9

FULL-COURT PRESSURE

Pookey's got Rasha pinned back—half out of her chair, shoved up against the wall. Her face terrified as he screams, "94 feet of hell! This is it, right here! Can't pass, can't dribble! Where you gonna go, huh!?"

"Pookey . . . !?" wails Rasha.

"I body you up like this, where you gonna go!?"

Rasha tries to wheel away but Pookey bodies her up even tighter.

"Here comes the double team! You gonna split it? You ain't gonna split it, not when I got you like this!!!"

Pookey looks over at me and grins maniacally.

Sexual harassment? I think not. Pookey's the most respectful guy I know.

He's just out of his mind. And Rasha knew that when she signed on.

We're up in his office on Melrose Avenue. I need name players for *94 Feet of Hell*—LeBron, Carmelo . . . how about Luke Walton? That's an interesting call.

Pookey knows everybody. Hence my visit.

"God, there was one game," says Pookey, finally releasing Rasha from the full-court pressure, "playin' against Providence—Rick Pitino."

"When you played for Seton Hall," I say as Rasha recovers in her chair and rolls herself to safety.

"Yeah. They called me the press buster, right? You give me any full-court pressure, I will dribble and twist and squirm and bust my way through it. Then came Providence."

His face contorts at the memory. "*That* was 94 feet of hell. Nightmare pressure . . . Just gettin' the ball inbounded, right? Then when that trap comes, I swear, man, it felt like they had six, seven guys on the floor. I died out there. Turned the ball over, threw it away, dribbled off my foot. P.J. benches me. Finally it's halftime. In the locker room, I call my dad, 'cause we're on TV, right? And I say, 'You watching?'

"And he says, 'Stop thinking. Just go. Get the ball and go. It's what you do.'

"So I get myself settled down. Go back out there, thinkin' okay, my dad's right, I'm good, everything's good, and guess what happens?"

Pookey breaks out laughing.

"They killed me again! Slaughtered me! There was blood on the floor, man!"

Rasha says, "That's nice, Pookey, you calling your dad like that."

Pookey smiles wistfully.

"So, Rob, this movie you're doin'—if you can capture that, that's cool beans."

Pookey knows LeBron's agents. Pookey's tight with Shaq and Alonzo Mourning. Pookey produces "Chocolate Sundaes" at the Laugh Factory on Sunset and all sorts of Hollywood people swirl through his world.

I leave with the promise that he'll help me any way he can.

My life's turning into a roller coaster. I've got paper flying everywhere, a hard drive crammed with emails, producers to call, agents to meet, contracts to be negotiated.

But I'm prepared. I'm ready. You think all those games of solitaire were in vain? All those times that I could've thrown in the towel, walked away from the computer, abandoned the mouse like a quitter. But no, I persevered. I played until my eyes grew bleary, until my hand cramped and my deep veins clotted. I know what it means to hang in there.

Last week in my column I included that story about getting hassled by a cop on the Venice Beach walk. I got emails from a couple of cops. This stuff stings. They want to be understood, a few bad apples . . . you know the drill.

Look, this country is filled with maniacs. Alcoholic, drug-addicted, gun-toting, spreadsheet-cheating, antisocial time bombs of every stripe, color, religion, ethnicity, and social-economic status there is, all lurching, flailing, or lying in wait, just itching to go off on someone.

As citizens, every day we try to slide past these people so they don't mess us up. For some this means checking the hidden costs of their mutual funds. For others it means ducking bullets as they step outside their doors in the morning.

I once caught a comic on BET—talking about driving his date home, parking in front of her building. She asks, "Aincha gonna walk me to the door?" And the guy answers, "Yeah, baby, but then who's gonna walk me back to the car!?"

It's hairy out there. It's nuts.

We've got two million people locked up, which is insane in itself, 'cause half of them did nothing worse than buy drugs for their own personal use from a cop. (And they say drugs don't make you stupid?) But for this they're in prison learning how to be really really bad. Most of them are gonna get out. Plenty of them are out already. And I'm talking about the hard-core criminals too. The violent ones.

The numbers are frightening. According to the U.S. Justice Department Bureau of Statistics, in Los Angeles County there are enough parolees to fill the Rose Bowl. Over 100,000 mostly young men out on the streets after doing years of hard time. Nationwide, four and a half MILLION on parole or probation.

And who do we ask to protect us from these criminals? Cops. Good cops. The guys on the thin blue line.

What was the old saying? "Don't like cops? Next time you're in trouble, call a hippie."

The phone rings—it's Sports Zone 790 in Atlanta (or something like that). Oh, shit, I'd agreed to do a phone interview. Suddenly I'm on live to over 100,000 listeners (felons all, I'm sure). Ten frenetic minutes with two moronic sports jock Howard Stern wannabes, and their half of the interview goes something like this —"Who'd you bang? Who'd you see naked? Is Jamie Lee Curtis a man? Who'd you bang? Who'd you see naked? Who's a jerk? Woody can't jump, can he? Who'd you bang? Is Whoopi a jerk? Who'd you see naked? Who'd you bang? Who'd you bang?"

I hang up, take a long hot shower, and return to the computer. Another email—this time I'm getting called out for referring to *94 Feet of Hell* as a war movie.

"You're disrespecting the guys fighting the real war—bodies piling up in Iraq—just the way Kellen Winslow did."

I disagree. War is the ultimate human pressure cooker—everyone knows that. By comparing sports to war, you're granting respect for the soldiers fighting the real deal.

Here's a reality check—flying back with my wife and kids to D.C. at Christmas. Chance to spend a week with my mother-in-law (who my wife insists didn't mean anything when she gave me a book for Christmas, *My Life as a Fake*).

On the plane, my six-year-old spots a soldier—young, tired, red-eyed, his uniform stripped-down basic, his boots scuffed. To my son, this uniformed young man is a god (then again, so is Mussolini, and half the members of the Village People). But my kid's too shy to say hello. So he just stares.

Later, somewhere over Oklahoma, I'm standing in the rear of the plane, stretching my legs, keeping the blood flowing, when the soldier wanders back.

"Where you headin'?" I ask.

"Iraq. Dulles to Frankfurt. Then Baghdad."

A flight attendant looks over. The moment goes tender, because women have that way about them.

"How is it there?" she asks.

"It sucks," he says. "The people hate us. They hate each other. Now they got us working to train the new Iraqi army. These guys, they're walking away. They got a gripe, they walk away."

"How much time do you have?" I ask.

"Three months. I got a wife, man. We wanna start a family. But now, with this stop-loss thing, they can keep me forever."

According to the *Washington Post* (December 29, 2003), the new stop-loss order can retain active military and reservists for years past their contracts. Army spokesman Maj. Steve Stover said, "We're all soldiers. We go where we're told. 'Fair' has nothing to do with it."

The flight attendant and I wish the young soldier luck. I can think of a few chicken hawks in Washington who ought to be reading *My Life as a Fake.* I don't know about you all, but for me, if we're gonna send our young people off to war, let's have someone making the decision who's been there himself.

Enough. Stay focused. You've got a movie to produce.

I scan through my emails. A writer friend—this guy has done it all: novels, journalism, movies, TV (good TV). This guy writes like there's no tomorrow. This guy's as masochistic as a writer can be. How masochistic is he? Season tickets to the Clippers.

His email reads:

"Rob, with this column, you have power. Be not afraid to use thy power, O Rob! Smite them mighty blows with thy terrible swift pen!

"Will they still hire you? Yes. Why? Because they'll be afraid not to! Before, you were just another writer. Now you can crush them in print! They can't afford to offend you!

"Rob, be not afraid. Do not grovel. Rather, crush! Kill! Maim! Destroy, O Rob! Plunder! Pillage!

"Your Humble Servant, B . . ."

I email him back, "I don't want to pillage, man. I wanna get a friggin' movie made. By the way, I'm putting your email in next week's column, can I use your name?"

"Hell no," he answers.

10

IT'S THE DISTRIBUTOR, STUPID

Wanna make a movie? Lesson No. 1: It's the distributor, stupid.

This country is littered with the rotting corpses of unseen movies—all financed on the backs of Mom and Dad, Uncle Joe and Aunt Rose, Capital One, Providian, Cross Country Bank, a consortium of dentists from Omaha, a consortium of drug dealers from Boca Raton, and undoubtedly a consortium of drug-dealing dentists from somewhere in between.

Don't get me wrong. Finding money to make a movie is no stroll in the park.

I've got an old teammate, John Hummer, who became one of the hottest high-tech venture capitalists in the world. I've been trying to squeeze money out of Hummer for years—all the way back to when he was playing NBA ball with the Buffalo Braves. No luck so far, although a business plan to launch that four-on-four full-court summer pro basketball league still holds great promise. (We want to give these guys the room they need to really play the game.)

But in Hummer's words, "The longest distance in the world is from an investor's pen to his open checkbook."

Still, people do write those checks. Then they can't get their calls returned. Because finding a company to distribute your movie is even tougher than financing it. Baron Davis, who played at UCLA and is now tearing up the NBA, played it smart. His movie, *Asylum*, is being distributed internationally by Momentum Pictures this coming fall.

Maybe it's the L.A. connection. In fact, Baron was at the very first four-on-four trial game we ran in Pauley Pavilion.

Anyway, without distribution, you're dead. Why start if you can't finish?

But, nonetheless, I'm convinced I've got a good shot at pulling off *94 Feet of Hell*.

For starters, it's a unique concept—the first sports movie that's all about one game, with no flashbacks, no setting the scene, no corny love interest to bog things down.

It's also the first sports movie where it isn't obvious which team's gonna come from behind and win at the buzzer.

In fact, I'm thinking of shooting two endings.

Now suddenly a real-deal producer has come forward and offered his services. He found me through my column. He wishes to remain anonymous. I'll call him J.C. We're in that due diligence stage right now.

What's "due diligence," you ask? Well, you know when you're walking your dog, and another dog walks up, and they start sniffing each other's assholes? That's due diligence.

I'd like to direct it, but would gladly step aside in order to get it made. There's Ernest Dickerson—he's been Spike Lee's director of photography. He's also directed one of the hippest sports movies ever: *Big Shot: Confessions of a Campus Bookie*.

There are other directors out there as well.

In the meantime, we're going to need the right arena. Time to bring on a location manager, right? Sorry, no budget yet.

But, hey, I've done some location work myself. It's one of the coolest jobs in the movie business. Running all over the place, snapping pictures, making deals. Especially if it's New York City. The year was 1978. (And for all of you wondering, God, just how old is this guy? I've got two words of advice when it comes to your own lives: Don't blink.)

It was a very weird time. See, I came of age with the Four Tops and The Stones. Jimi Hendrix. Neil Young. The Allman Brothers. Led Zeppelin. Van Morrison.

Barefoot, brown-eyed girls wearing string T-shirts and cutoff jeans—that's what worked for me.

But here I was, in the heart of New York City, having to face reality. The dream was over. It was time for . . .

DISCO!

Donna Summer and John Freakin' Travolta. Platform shoes and high heels. Oh, my God.

Dude, where's my culture?

I was living the life of a struggling writer in a crappy walk-up on Second Avenue.

I'd already worked on two movies, including a super-low-budget slasher movie (never released, of course) called *Heritage of Blood*, which we renamed *Heritage of Egg Salad* because that's what we got for lunch EVERY FRIGGIN' DAY!

The other movie was *The Bottom Line*, aka *The Death Collector*, aka *The Collector*, aka *The Enforcer*, aka *Family Enforcer*. It featured the brilliant Joe Pesci, who later starred as De Niro's brother in *Raging Bull* as well as that annoying guy ("Okay, okay, okay, okay . . . ") in those endless *Lethal Weapon* flicks.

Anyway, I get a call. It's John Stark, who'd been the production manager on *Death Collector* and would later marry and divorce Glenn "*Fatal Attraction*" Close (and somehow live to tell about it).

A big Hollywood movie (to us) was rolling into town—*The Warriors*—and Stark needed production assistants. On *Death Collector*, I made $70 a week. *Warriors* would be paying $50 a day! I jumped at the opportunity.

The movie was based on a gritty book by Sol Yurick—a real hard-nosed take on gang life in New York. Walter Hill would come on and (in my cinema-verité mindset) turn it into a cartoon. But, oh, what a cartoon.

Why do certain movies resonate in the collective psyche?

Who cares? Here's what happened.

I show up at the production office for the first day of work, and *boom!* the shit starts hitting the fan. Walter Hill has fired his first assistant director, and they're flying a new guy in from Hollywood. The producer, Larry Gordon, is ceaselessly dumping on Joel Silver, the young associate producer (who would later turn the tables by becoming the screamingest guy in Hollywood). Then there's the executive producer, Frank Marshall, grinning his way through the whole chaotic scene.

These guys (young Turks all) went on to produce a slew of famous movies—the *Lethal Weapon* series, *48 Hrs.*, *The Matrix*, *Raiders of the Lost Ark*, *Seabiscuit*, *Who Framed Roger Rabbit*, *Back to the Future*, *Alien*, *Die Hard*, and *Field of Dreams*, among many, many others. An incredible body of work.

And coming eventually, *You'll Never Die in this Town Again*.

God, it was rough-and-tumble. Walter had a John Ford-Sam Peckinpah thing for testing his actors' toughness. We had teamsters hanging out, drinking pints of whiskey and brawling with each other. Paramount was pissed about accounting irregularities. There was paranoia, anger, jealousy. A typical Hollywood movie, I figured.

I befriended a couple of our security guys—off-duty NYPD detectives. Big, burly Irish guys in black raincoats who I personally witnessed beat three onlookers to bloody, whimpering pulps because they mouthed off to them.

I was sliding through this mess, trying to keep my head down, when Frank Marshall started needling me.

"Hey, Princeton. Princeton fairy. How's my Princeton fairy today?"

Jeez, why do grown men behave like this?

I felt like saying, "Hey, Frank, you think you're gonna get to me? I played basketball in the Palestra down in Philly with 8,000 maniacs screaming, heckling, throwing coins and batteries, booing and spitting at us; and you think I'm gonna let some wuss Hollywood producer get to me?"

But did I say it? No. I was making $50 a day.

But I hated it. I was one of a dozen P.A.s. Bottom of the production. Everybody dumped on us. Then the production manager, John Stark, pulled me aside.

"Listen, Walter's really ticked about our locations. We don't have a place to stage the big gang meeting; we don't have the street to blow up the car; we haven't found a bathroom to stage the subway brawl in. I'm moving you up to locations."

"Yeah?"

"Yeah."

"Uh, more money?"

Stark said, "Hey, fuck you, you want it or not?"

"I'll take it, man, I'll take it."

"Okay. But you better come through for me."

I'm on the crosstown bus home late that night, feeling, Okay, I caught a break here, a chance to show these guys what I can do. I'm not a grunt anymore. I am somebody. I am a locations scout on a big-time Hollywood movie.

At dawn the next morning, over at Columbus Circle, they don't have a car for me yet. So I have to ride in the van.

I climb in the front seat, feeling like God's gift, and stick my hand out to the teamster behind the wheel.

"Mornin'. I'm Rob Ryder."

"So what?" he says.

HAVE A TAKE, DON'T SUCK

By that afternoon, I had my own car. Assignment—find a bathroom for the subway brawl between the Warriors and the Punks. ASAP.

So I hit every high school, every seedy hotel, every YMCA, hospital, movie theater, and public park, from the Village to the Bronx, Queens to Hell's Kitchen. I'd double-park, rush up, Polaroid camera in hand, find the bathroom, hold my breath, and step in to check if the layout was right.

(And who said movie work wasn't glamorous?)

Pity the poor bastard just looking for a few moments of peace and quiet, taking a dump, reading the sports section when suddenly there's this lunatic shouting over the stalls.

"Hey, everybody, I work for Paramount. We're making a movie! So just sit tight and keep the doors closed while I take some pictures, all right?"

"Que?"

"Uh . . . you know, movie, uh, cinema . . . ?"

Until finally I just started walking in and shooting, boom, boom boom boom boom. Five quick shots—doing the panorama—then I'd run back out to the car, where I'd tape them together and rush them over to that day's location, looking to catch the director, Walter Hill, at a good moment to present my latest discovery.

"Too small," he'd say.

And I'd spin on my heel and head back into the teeming fray.

"Too big."

"Too many windows."

"Not enough windows."

This went on for three grueling days, and I could tell that Walter was taking great delight in my increasing frustration.

Finally I worked up my nerve and said, "Are you fucking with me?"

Walter laughed.

"Not you. Paramount . . . I'm gonna make 'em build me a bathroom."

Ah. So this is how it's done. Paramount was tightening the purse strings and Walter was fighting back (in classic passive-aggressive style). I was just some flunky stuck in the middle.

I called Walter a couple days ago to reminisce. He said, "You know, by them not giving me more money to make that movie, I think it turned out better. There's a simple quality to it that I would've missed with a bigger budget."

Interesting concept.

Here's another example—John Landis directs one of the funniest movies ever, *Animal House*, for like three dollars and 11 cents. I mean, that was low-low-budget. And it turned out sloppy and gutsy and raunchy and full of energy and became a huge hit.

Two years later, he makes *The Blues Brothers* and of course he's riding high, and Universal's throwing a lot of money at him, and he takes a pretty cool, funkadelic scenario and screws it up with endless car crashes that took you right out of the movie. Why'd he do it? Because he had too much money.

Not my problem with *94 Feet of Hell*. I'm planning on doing this on a shoestring. (Under three million dollars.) Over 30 million people watch the NCAA championship every March. That's my audience. But I don't expect they'll all be willing to spend eight bucks at a movie theater, so I'm determined to make *94 Feet* lean and mean.

One of the biggest expenses shooting a sports movie—extras. All those people in the stands.

I've got to find the perfect arena (locations again)—small enough to fill without breaking the bank, but big enough to look real deal.

Billy Friedkin had a stroke of genius on *Blue Chips*.

Let's see, can I re-create the moment without self-inducing a paranoid paroxysm of putrid pustules?

In a production trailer on the Paramount lot. My buddy K.B. and I were working the phones, wrangling the basketball players. Shaq had already signed on. I was chasing down Chris Webber (who was being repped by his aunt) while K.B. was chasing Penny Hardaway.

I failed. K.B. succeeded. And Shaq and Penny ended up getting tight during rehearsals. (And for all you Orlando Magic fans who died when you heard that Orlando was trading the studly Webber for the softie Hardaway—blame K.B.)

Anyway, Friedkin comes storming into the packed production trailer and we all duck, because with Billy, you never know when he's gonna go off. But now, he's exultant.

"We're goin' to Indiana. Fuck this extra shit! We'll make *them* pay!"

And we did. The locations guy found the perfect 6,000-seat high school arena in Frankfort, Indiana, and for two straight days, those good citizens of the Hoosier state *paid* to be extras.

Three bucks a ticket. A stroke of genius.

It didn't hurt that Friedkin was tight with Bobby Knight and Knight had agreed to bring in a bunch of his former players for two key games.

Friedkin and Knight. Between the two of them, they had cornered the market on the world's supply of venom. Plus they liked to mix it up, these two.

Let's see, uh, *Godzilla vs. Rodan*, nah.

Frankenstein vs. Dracula—closer.

Alien vs. Predator—now we're getting somewhere. . . .

A sports agent named Mike Higgins (worked for David Falk, MJ's old agent) told me about a night out at a Bloomington restaurant riedkin and Knight started trading insults (it was a homosexual th as Mike recalled) and it got nastier and nastier—the two of them snarling, virtually spitting at each other (the whole restaurant going silent) and Mike slowly sliding under the table thinking, Jesus H., is this for real?

Until finally these two grown men, giants in their respective fields, burst into laughter. And the good churchgoing people of Indiana laughed with them.

Freddy vs. Jason—that's it, with a dash of *Chucky* thrown in.

Make the extras pay. Can I pull this off? We'll see. Certainly not in Southern California.

Meanwhile, I grab breakfast with Ron Shelton to talk about the script.

"The key is to show the game you don't see as a spectator. The words you don't hear."

An inside look. That's what I've written. Of course, it needs some tweaking, but if there's one thing I've learned about writing, it's this: Don't let people pull you in too many directions.

I'm not real big on all these seminars and technique books, but as long as I'm lurching about (and in response to a number of email inquiries), here are a few tricks of the writing trade.

First—(in the words of sports radio jock Jim Rome, no less): "Have a take, don't suck."

Second—"Make sure something interesting happens every ten pages."

Third—from the crime novelist Elmore Leonard (*Get Shorty*; *Out of Sight*): "Leave out the parts everyone's gonna skip over."

Anyone still with me?

Look, who wouldn't want to be a writer? You get to set your own hours, imbibe at will, and say what other people only think.

More advice—learn a trade where you can work in bursts. Me, I became a carpenter and then a movie production guy (my father rolling in his grave, asking, "For this I paid how much tuition?").

And most important, develop a thick skin.

After I finished my first screenplay (which I was convinced was an absolute work of genius), I approached this writer I'd met, Malcolm Braly, to give it a read.

Malcolm (R.I.P.) was a middle-aged ex-con who'd written a solid novel called *On The Yard* which later got made into a movie.

Malcolm split his time between CBGB in lower Manhattan and the Catskills, where he was hanging out in a farmhouse with a bunch of slinky women with snake tattoos running up their thighs and some pot-smoking, wine-guzzling, long-haired tough guys, most of whom had been in 'Nam or the joint and weren't quite ready to get with the program.

Malcolm was a sweetheart of a guy, though. He read my first screenplay, sat me down, and said,

"At least now we know you're not a genius."

Hey, man, I asked for it. Thick skin.

Where was I? Oh, yeah, New York City, 1978.

So Paramount finally agreed to build Walter's bathroom, most of the other locations had been secured, and I was free to hang around the set.

I knew Walter was a sports junkie, so I let him know that I'd played some college hoops. He liked that. He also liked (for some odd reason) that I'd led the Ivy League in fouls. And one day, he asked me, "How tough are you?"

"Uh . . . Jesus, I don't know," I said.

Wrong answer. When someone asks you how tough you are, the answer is "Tough enough."

"No, really," said Walter. "Can you handle yourself? Can you handle a bat fight?"

"Yeah. Sure."

"We lost one of our stuntmen. Baseball Fury. You wanna be in my movie?"

12

LEARNING TO FIGHT

The next night, I found myself down in Riverside Park with a 36-inch Louisville Slugger in my hand. Wearing a knockoff Yankees uniform, black cap, black wig, my face painted a weird black and purple. Just like that I went from a $50-a-day locations scout to playing a Baseball Fury, raking in the Screen Actors Guild minimum of $300 for the first eight hours. Plus time and a half overtime, and in the movies, there's always overtime. 12, 14-hour days (or nights) are not uncommon.

(Mr. Hill was to come through for me twice again once I got to Hollywood. Early on, he used a line on me that I continue to use to this day—"The least we can all do is lie for each other.")

But Craig Baxley, the *Warriors* stunt coordinator, was worried. "Don't rush it! You rush it, you're gonna screw up the sequence and someone's gonna get hit in the head."

I was about to face off with Swan, the warlord of the Warriors. He'd taken a bat off one of the other Baseball Furies and I was the last obstacle to his escape.

You're not alone if you consider *The Warriors* to be one of the all-time great tough-guy movies, but let's be real—most of the actors were wimps.

Short, skinny. Soft. Pouty lips and poofy hair.

There were two exceptions. First, Michael Beck, who was playing Swan. He was muscled up, an obvious athlete, with a quiet, steely intelligence. Then there was James Remar, playing Ajax. Remar had big biceps and an uncontrollable sneer and was an obvious lunatic.

The Warriors wore those open, sleeveless leather vests, and before each take Beck and Remar would knock out a bunch of push-ups to pump up their arms.

At first, the other Warriors had tried the same thing, but everyone laughed and they gave it up. I was glad to be facing off against Michael Beck instead of James Remar. Remar delivered that memorable line—"I'm gonna take this bat and shove it up your ass and turn you into a Popsicle."

He would go on to play a whole string of tough-guy roles.

While Michael Beck would next play the disco-roller-skating love interest to Olivia Newton-John in one of the worst movies ever made, *Xanadu* (also produced by Larry Gordon).

As the *Kalamazoo Gazette* put it about *Xanadu*, "There's never been a movie quite like this, and there never will be again. And that may not be a bad thing."

It was the beginning of the end for Beck, who entered that long, slow skid of diminishing roles that is the fate of so many actors. So close . . . so close . . .

But there we were. Two young bucks. Mano a mano. Bats in hand.

Ah yes, one more Hollywood depiction of mankind's genetic predisposition to beat the shit out of each other.

It's pitiful, isn't it?

You know what gets me? This whole Marquis of Queensberry, rules of engagement, Geneva Convention bullshit.

Let's see—how about the use of mustard gas that scars your throat, burns your lungs, and poisons your blood, causing paralysis, convulsive vomiting, internal bleeding, severe blistering, bulging eyes, excruciating pain, and near-certain death?

Nah, too nasty. Against the rules.

Then will you let us use cluster bombs? These devices of doom contain hundreds of bomblets which explode with volcanic percussive force, propelling thousands of sharp shards of steel that indiscriminately shred human tissue and bone, slicing through arteries, bursting vital organs, causing excruciating pain and near-certain death (provided they're not the ones that remain unexploded, only to go off years later in the hands of children who are attracted to their bright colors and interesting shapes).

Yeah sure, why not? As long as we can use them too.

It's unbelievable. Rules of war.

You know where the phrase "rule of thumb" comes from? An actual law once on the books in England that you couldn't beat your wife with a cane that was thicker than your thumb.

Yet three of the great moments in movie history come from a character's breaking the rules.

Indiana Jones, facing off with a turbaned Arab who performs this elaborate, menacing sword demonstration. Remember? Indy smirks, pulls out a revolver, and shoots him dead.

Then Butch Cassidy (the diminutive Paul Newman) gets challenged by Harvey Logan, a fellow Hole-in-the-Wall gang member (the towering, tough-as-nails Ted Cassidy).

They're about to fight over who leads the gang. Butch is hesitant, shuffling up to Harvey, who's looming there, bare-chested, legs spread, a huge hunting knife in his hand, egging him on.

Butch says, "No, no, not yet. Not till we get the rules straightened out."

The incredulous giant goes, "Rules?! In a knife fight!? What rules!?"

At which point Butch Cassidy kicks him in the nuts and the fight is over.

Then there's the incomparable opening of 2001: *A Space Odyssey*.

Remember that human-ape picking up a mastodon bone and feeling its heft? You could actually sense his primitive brain thinking, Man, I could do some serious damage with this baby.

Moments later he whomps a rival on the head, so beginning an arms race that continues until this day. There's no other species on the planet that does so much damage to its own members. We're doomed! Doomed! But, hey, might as well have some fun with it.

My wife's the director of a primary school. It's a job that's not unlike directing movies, which, as the wonderful director Sydney Pollack puts it, "Is a lot like getting pecked to death by ducks."

Anyway, her school provides a warm, nurturing environment. When a child gets hit on the playground, there's no hitting back, it's "Go find a teacher."

Just recently, our six-year-old, Andre Xavier Ryder, complains to me about getting bullied. (So his name may enable us to hustle a scholarship off some unwitting college basketball coach, so what?)

I pull Andre aside. "Next time it happens, give the kid a hard shove in the chest. *Then* go find a teacher."

"Weally, Wob?" he asks. Andre stopped calling me Dad the day my wife took our older son off to Paris for two weeks. And he's still struggling with his Rs. "Wob. Wob Wyder. Wobby Wyder. Wobby William."

"Yeah," I say. "Shove him back, *then* go find a teacher, AND DON'T YOU DARE TELL MOMMY."

Every man carries a moment when he should've shoved back. Even if it meant a certain beating. A junior high school humiliation. That time on the bus. In the locker room.

The ones who don't fight back—especially the intelligent ones—they internalize these humiliations where they fester into gnarly balls of hate, lying in wait until they finally erupt as all sorts of twisted mind games and verbal terror on some defenseless underling in a never-satiated cycle of eighth-grade revenge.

Hollywood is filled with men who should've shoved back when they were boys.

Am I wrong to advocate physical defense?

Here's what I taught my 12-year-old, Cole: "If push comes to shove, throw two fast left jabs to the face, followed by a hard right to the body, then tackle the kid and tie him up and hope someone comes save your ass before he bites your ear off or his friends stomp you."

Rules of engagement. Cole and I practice so he'll be prepared.

Sitting around at dinner with my little nuclear family—my wife, Andrea, sons Cole and Andre, and me. My wife puts on that stern face as she addresses our six-year-old.

"Andre, Ms. B. said you pushed Jackson today."

I eyeball my kid—Don't you dare rat me out.

"Andre?" his mother asks.

I keep staring at him, my eyes growing wider.

"Andre!"

The poor kid stares imploringly at me, then turns to his mother.

"Wob didn't tell me to."

13

CRAZY WHITE PEOPLE

My guys at ESPN.com didn't even call. Tuesday morning I got to Page 2 to read my latest column, and it's not there. Uh-oh. I click over to the newly created Page 3, and there I am—Hollywood Jock. What's Page 3, you ask? Let's put it this way—if Page 2 is for jocks, Page 3 is for jock-sniffers. Page 3 is another attempt by ESPN to marry sports and entertainment. My fears of being moved were finally realized. And now, instead of writing alongside guys like Hunter S. Thompson and Ralph Wiley (both now dead, btw) my new colleagues on Page 3 will be writing groundbreaking articles about who Carmen Electra thinks has the sexiest legs in the NBA.

Plus, my weekly readership will immediately plummet (from a high of 140,000-plus to a new average of 20,000 or so). Aw, man. Aw, fuck.

I call my old Page 2 editor, Michael Knisley. Knisley's been a real prince since day one, and I'm gonna miss him. Knisley's apologetic. He fought to keep me on Page 2 for as long as he could, telling me finally, "Look, I decided to stop arguing with my boss."

Okay, all right. It's a done deal. The column's still alive. I've got bigger fish to fry. And as long as I'm now a certified member of the celebrity bullshit school of journalism, how about this:

NBA All-Star Weekend, the biggest African-American block party of the year, is fast approaching. My pal Pookey Wigington is throwing a huge Chocolate Sundaes All-Star party Thursday night at the Avalon in Hollywood. He wants me to be there.

"We got Serena Williams hosting. We're gonna have Bill Bellamy, Tommy Davidson. . . . It's gonna be huge, man. Fourteen hundred people, all sorts of surprise celebrities . . ."

It's one chance among many this week for me to hook up with some of the NBA players and agents I'm trying to attract to *94 Feet* (amended from *94 Feet of Hell* due to overwhelming demand).

The NBA All-Star game has become a commercial juggernaut.

TNT's just getting a grip on this golden goose they've bagged. Last year, for the first time ever, the game was on cable and it scored HUGE ratings. Shocked everybody. And corporate sponsors have jumped into the mix big time.

The American Express Magic Johnson All-Star Celebration.

The McDonald's Mascot Slam Dunk Contest.

The Radio Shack Shooting Stars Competition.

The Foot Locker Three-Point Shootout.

The Sprite Rising Stars Slam Dunk Competition.

And finally, the main event, the Gatorade Tip-Off NBA All-Star Game.

Talk about milking it. Oh, yeah, I forgot,

The got milk? Rookie Challenge.

It's Friday the 13th, almost midnight, NBA All-Star Weekend, and things are looking dicey. I'm alone behind the wheel of my wife's little white minivan, stuck between a gigantic black Hummer and a looming black Escalade (whose rims alone could finance four years at USC).

Santa Monica Boulevard is all jammed up, everybody trying to make the right turn onto Formosa, where the valet parkers scramble to handle the onslaught.

This is the Reebok party, hosted by New Orleans Hornet star Baron Davis (since traded to Golden State), and the word is out: This is *the* place to be—the old Warner Lot in Hollywood.

I'm frazzled. Working on three hours' sleep. I was up way late last night at the Chocolate Sundaes comedy-dance party at the Avalon. Chris Spencer and Pookey Wigington's party.

An impressive entrepreneurial effort. They'd managed to attract Serena Williams, Larenz Tate, Tommy Davidson, Ludacris. They'd charged 30 bucks a head, providing a chance for all those people who don't know the right people to still go to a big All-Star party.

Some paid 60 bucks for the VIP ticket, which got you a blue wristband that entitled you to wave it around, letting everyone know you were a VIP but not much else.

Me, I got the yellow wristband, care of Rasha and Maya (Pookey's slaves, as they refer to themselves), which identified me as a VVIP (Very Very Important Person) and entitled me to some chicken wings and carrots and celery in a private upstairs room.

No free drinks at Pookey's party. These guys were looking to make some money while showing the salt of the earth a real good time, and you can't help but admire that. The American way.

But tonight, right now, this is different. This is Reebok. This is a multibillion-dollar corporation doing the serious bling-bling ka-ching thing.

You don't *pay* somebody to get into this party. You *are* somebody.

We inch forward. At this rate I'm gonna arrive just in time to say good night. I see a break to my right, gun the van, and end up over on LaBrea—plenty of empty meters. I jump out, lock up, and stride toward "The Lot."

It's chaos out front. A swelling crowd of twenty-somethings, mostly black, dressed to the nines, clamoring to get in. Still others streaming out of the Hummers, Escalades, and several ridiculous 30-foot-long SUV limos that look right out of Mel Brooks's *Spaceballs.*

I pick up my pace, putting on my gate-crashing game face.

There's a trio of security guards stopping people who've shared my bright idea—a flanking action from a side street.

I don't try to duck them, instead heading for the biggest guy.

"Where's Andy!?" I shout.

"Say what!?" he barks as he moves to block my path. "You can't—"

"You don't know? Jesus!" I feint right, then cut left and power past him.

"Hey!" he yells.

"I'll let him know where to find you!" I shout over my shoulder. "Just stay right here!"

He hesitates, looks confused.

Remember when Gene Hackman shook down that petty drug dealer in *The French Connection?*

"Do you pick your toes in Poughkeepsie?!" he kept shouting. "You do, doncha? You pick your toes in Poughkeepsie!" Until the guy just broke down out of sheer confusion.

Same idea.

There's a long bulging line of anxious partygoers waiting to get checked through by three handsome women wearing headsets and sharp suits and sporting those all-important guest-list clipboards.

I slip up behind one. "I'm looking for Andy Gelb. PMK? He said to meet him here but there's no way." She looks me up and down. "I'm with ESPN," I add.

She considers, then slips me a playing card—the King of Clubs, and I feel (but don't show) a flush of relief as the Jesse Jackson mantra comes to mind:

"I am!"

"I am!"

"Some-body!"

"Some-body!"

The soundstage is huge—dark and bright—a grid of swirling disco lights hanging from the towering ceiling. The music deafening, the bass line microwaving my internal organs.

A pretty girl with a tray of colorful vodka cocktails dances over and offers up. I take a couple gulps and survey the scene.

The enormous space is filling fast. There's the glare of television lights—several camera crews working the room. A dancer on stilts—an easy twelve feet tall. Photographers everywhere. Waiters swirling about with all sorts of canapés and hors d'oeuvres.

Four full bars strategically located. Free drinks. Quality liquor.

I spot Dennis Rodman across the room and my lip curls. Rodman worked with us on *Eddie* for a couple of days back in North Carolina, and he was incredibly insulting to me, and I'm still pissed about it, but I decide to let it go (for the moment).

I turn and spot John Salley and work my way over to say hello. He's smoking a long cigar and laughing and flipping people off and being the general all-around good guy John Salley is.

A sudden commotion—LeBron James makes his entrance, surrounded by his agents, Aaron and Eric Goodwin, and a whole bunch of others who have his (and their own) best interests at heart.

I decide not to approach the Goodwins tonight, even though Eric's read and "got" my college basketball movie, *94 Feet*. Another time.

I hug Salley good-bye, work my way to another bar, grab another cocktail, then plop myself down in a plush armchair in a far corner.

There's a mixed crowd sitting around on sofas, armchairs. An older black couple. A few hip-hop artists. Two scantily clad white girls with dollar signs in their eyes.

Another commotion. Baron Davis makes his entrance—looking sharp in a suit and silk shirt.

I know Baron a bit from his UCLA days. And just last week, I was in touch with his young agent (and former teammate) Todd Ramasar, about Baron's being in *94 Feet*.

Todd and Baron have put in place a smart business plan for a boutique sports agency.

Baron works his way through the crowd, laughing, hugging, shouting hellos.

I sit tight as it slowly dawns on me—I'm in a power corner.

Some impressive, all-business security types suddenly show up with portable posts and velvet ropes and begin roping off the corner.

My corner. I'm on the inside. Now this is a first. By sheer luck, I'm suddenly a VVVIP. And finally, there he is, stepping through the ropes, Baron Davis, offering me a big smile, a hearty hug. Baron shouts that he's heard about the movie, wants to discuss it when the time is right.

This is good news. See, Baron Davis is not just another baller looking to dabble in Hollywood. He's established a true film company, with seasoned partners and lines of financing and distribution.

In my mind, I'm making a backdoor play. "Hey, Baron, wanna be in my college hoops movie? Yeah, cool. And by the way, how about financing the thing while you're at it?" (Actually, his presence alone will garner serious interest.)

Baron moves off. It's time to step out beyond the velvet ropes.

The place is jammed by now—dance floor filled with writhing bodies. Beautiful women, good-looking guys— bumping and grinding to Missy Elliott and OutKast.

Suddenly, four gymnasts come flying down from the rafters on bungee cords. The crowd freaks as the gymnasts nearly splat on the floor before bouncing back up into the air, where they go into a series of flips and spins in sync with the throbbing music.

People are laughing and shaking their heads and you can just hear them saying, "Crazy white people . . . we're trying to dance here!"

I spot Duane Martin talking with Damon Wayans and I head their way. These two are married to the same woman, Tisha Campbell. How's that? Duane in real life. Damon in the TV show *My Wife and Kids*.

I can just imagine what they're talking about:

"You know what really gets me? When she starts bitchin' about the toilet seat . . . "

"I hear you, man."

Duane Martin I know from the Hollywood Y and *White Men Can't Jump*. He came out of Brooklyn and was a real baller. He's done well for himself, a string of movie and TV roles, and now he's starring in the UPN comedy *All of Us*, which is produced by Will Smith and Jada Pinkett.

Damon Wayans I know from *Celtic Pride*. The producers tried to hire me as the basketball adviser but I'd already committed to *Eddie*, so I offered up the services of my brother-in-crime Kevin (K.B.) Benton.

(What are the odds of two crappy basketball movies being made at the same time? Pretty good, given the ways of Hollywood.)

Anyway, *Eddie* had wrapped and I was back home in L.A. when I got a call from Boston.

It was the line producer. They were looking for additional help with their big basketball finale. After much pressing, he assured me I would be merely augmenting, not replacing, K.B.

I managed to squeeze a promise of 6,000 bucks and first-class airfare out of him for seven days' work, so off I went.

But Damon Wayans wasn't particularly happy when I arrived. It was a black thing. I couldn't blame him for giving me the cold shoulder.

And now, as I work my way toward him and Duane Martin, I'm thinking, Play Duane, not Damon.

And that's what I do. Damon turns away (probably not even recognizing me) as Duane and I shake hands and bump shoulders and I scream in his ear that I'm really glad how things are going for him and that some cool things are happening for me too and we should hook up soon, and he screams back that he'd be glad to.

I peel off before Damon can turn around and say, "Hey, you're the asshole who came in late on *Celtic Pride* and started telling everybody what to do."

I shove my way along, trying not to spill my drink while appreciating the press of some of the sexiest women on the planet who are heading in the opposite direction and have given up any semblance of modesty in the crush of bodies.

There's a second power corner that's developed across the way. A thick pack of people, and two of the largest bodyguards I've ever seen. One guy black, the other . . . I don't know what he is . . . some mix—part Samoan, part Hulk, I can't tell, but these guys are both an easy six foot nine, 380.

No exaggeration.

The Samoahulkian has that telltale bulging forehead that says, you just breathe this guy's exhaust you're not gonna be able to pass a steroids test for weeks.

The bodyguards' assignment—make sure P. Diddy and Jay-Z (two of the richest hip-hoppers in the business) enjoy themselves and get out in one piece.

I get as close as I can before the Samoahulkian starts eyeballing me. I consider –what am I gonna do if I do actually get there, introduce myself, and ask for money?

So I turn and work my way onto the dance floor, where by now total anarchy reigns—which means I can dance away the rest of the night without anyone mistaking me for Mark Madsen.

Later, I'll tell my sleepy wife, "Honey, you'd be proud—I danced with some beautiful women," and she'll answer, "Yeah, but did *they* know that?"

Suddenly there's Dennis Rodman again, leaning up against a wall across the way.

I don't know what does it, maybe the fourth vodka cocktail, maybe the steroid fumes coming off of P. Diddy's Neanderthal bodyguard, but suddenly I'm feeling belligerent.

14

BIRDS AND WORMS

Against all rules of self-preservation, I head Rodman's way, looking for payback.

I try to talk myself out of it: Don't do this, man. This is stupid. You've had a great time, it's late, everybody's drunk and dancing. Go home. Go to bed.

But, no. Not me. I'm from New Jersey.

It's that time in the booze-soaked party where people start falling out. That's the thing about open bars with quality liquor: You have to drink. It's like walking through a field of money. You gotta pick it up. You have a moral obligation to get your fair share.

And just because you can't walk out with a bottle in your hands doesn't mean you can't walk out with one in your stomach—ounces of alcohol coursing through your bloodstream—making you happy or sad, stupid or mad, or any combination of the above.

It's that time of night when you're sitting next to a guy on a couch and he slowly starts keeling over, eyes half-shut, his addled brain thinking, I know there's a cushion down here somewhere and if I just keep leaning over like this, I'm gonna get there 'cause it's there, I know I saw it. . . . I'm not in my car, am I?

Or the guy trying to cross the dance floor when he's had too much, and he'll do these flanking motions, like a sailboat tacking upwind, stumbling left for a while, then heading back right, like you improve your odds of finding your destination by covering more territory somehow—I'm gettin' there, I'm gettin' there. And maybe he'll stop to dance a few steps before some girl will give him a look of pure disdain that will send him reeling along on his way.

Whereas a drunk woman will cross the dance floor like she's on a mission. Walking that straight line—point A to point B—putting one foot in front of the other, steps short and quick, knowing as long as she keeps leaning forward and staying focused she's gonna finally get there.

Or another group of beautiful young women clustered around a girlfriend who's holding on to a column like salvation itself, eyes crossed, speech slurred as her smarter girlfriends argue.

"Who's taking her home? You think I'm takin' her home?"

"Well, I ain't takin' her home."

"You think I'm takin' her home?"

And me, moron that I am, heading for Rodman, a gnarly ball of alcohol-induced anger building behind my eyeballs.

So what did he do to cause such resentment?

He denied me my very existence.

It was on that Whoopi Goldberg-coaches-the-Knicks-movie *Eddie*, which I've trashed enough, so suffice to say, if you haven't seen it, don't.

Rodman was still playing for the San Antonio Spurs. We were in the thick of the schedule, shooting a whole bunch of basketball—multiple games: the Kings, the Hornets, the Suns—Kurt Rambis working his Rolodex, signing guys up, getting them in on time, and then essentially lying low in the office while I worked the floor, trying to get these millionaires to put some effort in for the cameras, but the ball came out soft because these guys were on vacation and this was a lark.

And Rodman?

He showed up like he didn't want to be there.

Ah, the dilemmas of the media slut.

I don't care if it's Martha Stewart or Ben Affleck, Paris, Michael, or Janet, these pitiful people with this pitiful need to keep thrusting their pitifully needy personas upon us is growing truly tiresome.

So the great Dennis Rodman had chosen to bless us with his presence on *Eddie*.

Only he didn't want anyone telling him what to do.

Correction, he didn't want *me* telling him what to do.

So I'd tell him through his agent at the time, Dwight Manley—this young squeaky-clean guy who didn't act like it was out of the ordinary for him to act as interpreter in a three-way conversation EVEN THOUGH WE ALL SPOKE ENGLISH!

It went like this—standing at midcourt, Rodman looking off into the rafters.

"Hey, Dwight, in this first shot, we'd like Dennis to come from the weak side, block Ostertag's layup off the backboard, grab the rebound, and outlet to Avery."

"Dennis, in this first shot, they'd like you to come from the weak side, block Ostertag's layup off the backboard, grab the rebound, and outlet to Avery."

"Uh-huh."

"Then, Dwight, we'd like Dennis to . . . "

It was ridiculous. But it was also disrespectful. And you know what? If you're thinking, Boo hoo hoo, just get over it, man, then you're not getting the point. Because when stuff like this doesn't get to you, *that's* when you have a problem. Because the day you become one of those sycophants who accepts this behavior is the day you surrender a piece of your manhood.

It wasn't the first time this had happened.

It was at the NBA Chicago rookie camp. I was looking for players—it was just before draft day—and I saw Larry Bird sitting alone, looking over this latest crop of NBA wannabes.

Since we'd both worked on *Blue Chips*, I walked up, stuck out my hand, and said, "Hi, Larry, I'm Rob Ryder, the guy who put the basketball players together for *Blue Chips*, and I just wanted to say hello."

He didn't say a word. Didn't move a muscle. Totally ignored me.

"Uh . . . Larry?"

He stared straight ahead. He was right next to me. I stood there feeling like a total idiot. There was no one within ten feet of us. He didn't turn and shake his head like, "This isn't a good time." He didn't say, "Look, I'm busy right now. . . . " He just sat there like I didn't exist.

And you know what? I didn't.

Finally I walked away, red-faced and pissed off.

Sportswriters get this treatment all the time. That's why things get so vitriolic. You don't think it makes these guys burn? I'm talking about the beat guys. The guys in the trenches.

But hey, no one made them become sportswriters, you might think. And you're right. You get what you ask for in this life.

What was that line from Ron Shelton's screenplay? (He told me it's a variation of a Bobby Knightism.) Something like, "Guys who can't play or write become sportswriters."

A totally unfair statement, as Ron readily admits. But hey, never let the truth get in the way of a good line.

And never underestimate the potential of writers to beat themselves up.

Masochists all.

You know who handles celebrity the right way? Jack Nicholson. "Just fucking say 'Hello,' that's all you gotta do." Thanks, Jack, for keeping it real.

Anyway, back with Rodman on *Eddie* in North Carolina. We were all ready for the first shot when the first assistant director rushed up and whispered to me, "He's gotta take his earrings out. And the nose ring."

"Um, sure, yeah," I said. "Hey, Dwight, would you please ask Dennis to lose the jewelry?"

"Dennis," said Dwight. "They need you to lose the jewelry."

So we all stood around, probably 180 people, cast and crew, camera operators, makeup, wardrobe, all of it, plus a couple thousand bored-out-of-their-minds extras as Dennis Rodman removed his multiple earrings.

But his nose ring was stuck. I could see he was struggling with it, but believe me, you don't wanna look too close.

The A.D. scurried back up and whispered in my ear, "What the fuck is goin' on?"

"Looks like his nose ring is stuck."

"Aw jeezus. I'll get a grip."

"Good idea," I lied.

So the A.D. rushed off and came back a minute later with one of the oldest, biggest, gnarliest, most white-haired, red-faced, New York–based movie grips I'd ever met. He was wearing a tool belt with about a thousand screwdrivers and pliers and wire cutters and God knows what all, and I was thinking, Aw man, how's this gonna go down?

So the grip walked right up to Rodman and said, "What's the problem, Dennis?"

"Nose ring," Dennis answered.

The grip stared up Rodman's nose like a dentist examining a cavity. "Yeah, okay, I gotcha."

He grabbed some needle-nosed pliers and a pair of wire cutters and went to work, and the conversation went like this:

"So, uh, Dennis, youse been seein' Madonna, huh?"

"Yeah. On and off."

"Yeah, I worked with her on *Truth or Dare.*"

"No shit."

"No shit. Nice girl, Madonna."

Then, snip, the nose ring was out, the A.D. yelled "Roll sound," and we were on our way.

That was 1995. A long time ago. And now I'm thinking, Let it go, man. It's late, you've had a great time, you made a nice connection with Baron Davis, so don't blow it now. Baron's the host. And he's got his own movie project in the works, *Blacktop*, and like me, he wants to use current NBA players in his movie.

As he recently told sportswriter David Aldridge, "Half the NBA acts anyway. They get their money and they act like thugs."

But no one's acting like a thug tonight. Everybody's been real cool, friendly, laughing, smiling. Rodman's standing with a couple of beautiful women. He spots me making my approach and tilts his head up and off to the side. Normally that's enough to get people veering away.

But not me. Not tonight. I walk right up to him, virtually toe to toe, living proof that alcohol is this culture's most dangerous drug. Three parts stupidity, two parts belligerence, and nary a twist of common sense.

"Hey, Dennis, how ya doin'?"

No response.

"We worked together on *Eddie*."

He stares off into space (which is still filled with bungee-jumping gymnasts doing somersaults to the banging music).

"Hey, Dennis, I said we worked together on *Eddie*. Remember? Just thought maybe you'd say 'Hello' or something."

He slowly turns toward me. He doesn't look angry, he's not filled with contempt. He just looks dull. And he actually says something.

"Long time ago, bro."

I slowly smile. I am now complete. Dennis Rodman has acknowledged my existence. I turn away and head back across the dance floor and only now do I experience a flood of relief that Rodman didn't do a Sprewell on me.

It's the small victories.

When you're not worth a million.

And nobody knows your name.

Chalk one up for the little guy.

Now where can I find Larry Bird?

15

IT'S FOURTH AND LONG

They call the William Morris Agency the black hole. Once you sign on as a client, you get your fifteen minutes of attention, maybe score a deal or two, then no one ever hears from you again. I was there years ago when I first got to Hollywood.

One day I'm in New York swinging a baseball at Swan in *The Warriors*, the next I'm in Beverly Hills sitting across from my new agent, a wonderful, matronly Brit named Judy Scott-Fox (R.I.P.).

Judy would send me off to meetings like a mother hen—"We don't want to be late now, do we?"

I'd written a spec comedy script about a small town next to a nuclear waste dump. Like everyone else, I was outraged by the nuclear power industry. Forget the danger—it's the hard costs that'll kill us. Our children's children will still be paying to decommission these Bechtel, fat-cat abominations. And now, with global warming all the rage, even some environmentalists are saying maybe we should build more. How moronic is that? All those sites are going to have to be maintained and monitored and secured for 10,000 years. I don't know about you, but my vision of the future is a lot closer to *Road Warrior* than *Please Don't Eat the Daisies*.

But like Judy said, "We shouldn't be too preachy now, should we?"

She found a terrific Englishman, John Irvin, to direct, and she optioned the script to a small production firm called MMA. My first deal. Of course, it turned to shit in about six months.

MMA was owned by an older husband-wife team who'd made their bread in the schmatta business and had a son who wanted to direct.

As the old man told me, "My son is either a genius or the biggest asshole on the planet."

They'd hired a guy named Steven Bach to run the company. Bach had just been fired from United Artists, which he'd virtually bankrupted by greenlighting *Heaven's Gates,* then refusing to pull the plug when the director, Michael Cimino, blew the budget up by doing stuff like shooting (and printing) fifty-five takes of a single moment.

Bach hired a development guy whose name escapes me, but like many a lower-level executive, he must've spent hours each night scrubbing off the footprints.

We were in a meeting addressing my script's "second-act problems," which is kinda like four guys wandering around in a swamp at night with a box of damp matches.

"Dude, Where's My Screenplay?"

Bach was pontificating about narrative confluence or something. John Irvin, the director, sat there thinking, Oh, God, here we go again. I was wondering if it was always gonna be like this. (It was.) And this poor sap of a development executive was desperately looking for a chance to break in with an idea, thereby justifying his miserable existence.

He finally got the chance and spat out a quick three sentences.

Total silence.

Bach turned to him and said, "That is the stupidest idea I ever heard."

I felt bad for the guy. Look, I played for a coach who once likened our best player to a dog turd he saw on the sidewalk that morning, so this stuff wasn't new to me. But still, it makes you wince.

Anyway, the option expired, I sold the thing once again, and I at least made a bunch of money, but like most scripts, it never got made.

The Writers Guild of America registered over 50,000 new scripts last year. This is what a Hollywood screenwriter is up against. You've gotta have several balls in the air, irons in the fire, pigs in the blanket.

Especially right now, because last night, my wife dropped a bomb on me. After 16 straight years as a teacher and principal, she needed a break. She's about to notify her board. Forget the fact that it's her paycheck that's been keeping the family afloat. Forget the health coverage for all four of us. Forget 401k contributions. She's done. I do some quick calculations. I'm already 10 weeks into my "Just one more year" scenario. And now I'm learning that in 24 short weeks, her steady paycheck will disappear. How's that gonna work? No way I'll ever go back to the credit cards. No way. That two grand a month from ESPN is gonna look pretty meager when it's the only deposit showing up in our bank account. I need a deal. At the very least, I need to get some cash flowing.

There's a Samuel Jackson hoops movie called *Coach Carter* shooting in L.A. right now. Word's out that they're dissatisfied with the basketball scenes, but I'm reluctant to make the call. These guys are pros, they'll get it fixed. At this point it'd be a step sideways for me. I gotta chase my dream. I gotta get one of my own scripts made.

Take Me to the River is stuck in an eddy. Pookey and Rasha both loved it, but so what? It's painful—here I'm sitting on a wonderful, funny, fish-out-of-water, interracial romantic comedy and it's slipping away into the mist. *Hoop de Ville* feels like a real long shot (although an extremely promising project). It's just that I don't know where to begin to find financing for a basketball musical. At the moment, my best bet feels like *94 Feet*. I need a director. There's an African-American, Rick Famuyiwa, who directed *The Wood* and *Brown Sugar*. Famuyiwa's a baller himself. I met him when a Pepperdine assistant coach, Wyking Jones, brought him down to Pauley Pavilion when we were running full-court four-on-four trial games.

So I put a call into Wyking. It just so happens that his boss at Pepperdine is former NBA player and coach Paul Westphal. And Westphal was just quoted in the *L.A. Times* saying he told David Stern to change the NBA game to four-on-four.

It's a great concept, way ahead of the curve, and I've invested a lot of time and effort into launching a 4-man summer pro league. But that too's gotta stay on the back burner for a while.

I ask Wyking to let Famuyiwa know I'm interested in discussing *94 Feet* with him. Packaging Rule No. 1—avoid agents and managers whenever possible.

There's another potential director, Paul Johansson, whom I met playing at the Hollywood Y. Johansson played on the Canadian national team, and he could flat-out play. Not quite NBA caliber, but a delight to run with.

Johansson has gotten gobs of acting work, particularly in TV. And he just directed Gena Rowlands in *The Incredible Mrs. Ritchie*. His old phone number's defunct, so I track down his manager through the Screen Actors Guild. Johansson's in North Carolina, acting in *One Tree Hill* for the WB.

The manager, Gordon Gilbertson, likes the concept, thinks Paul would be perfect to direct it. We riff on who's gonna win the NCAAs this year. He asks me to email him the script. I hang up and do so, thinking, He's never gonna read this.

Nobody reads.

To his credit, Gilbertson emails me back: "On 2nd thought just hold off for now! It's pilot season—so im crazy. it will be impossible to download at this time. thanks gg"

What has this world come to when a writer is appreciative when someone actually admits he won't read something?

I call my book agent, Matthew Guma, in New York. He's with Arthur Pine Associates. I figure the associates must be running the show since Arthur Pine is dead. They say he was a great guy and a great old-school agent.

Guma, my new-school agent, is a wild one. A former North Carolina Tarheel and a maniac for college hoops. He's convinced he can get me an advance on a novel I'm writing, so that's good. He's also just been to a Knicks game with LeBron's agent, Aaron Goodwin. Things are coming full circle, since I've spoken with Aaron's twin, Eric, about LeBron being in *94 Feet*.

What else?

I head into a Lifetime channel meeting with a writing buddy, J.B. White. We've come up with a cool series concept, *Holly & Vine*. Squeaky-clean actress teams up with hip-hop P.I. to solve cases in Hollywood's underbelly.

We've got a showrunner/producer/director attached (all through the William Morris Agency, which means that sometimes they *package*, they really do *package!*).

We pitch the show to a couple of African-American women who run the series division, and they seem receptive. We'll see. It's too cool an idea not to get made. Or at least not to garner a script commitment from Lifetime (which would mean a cool 40K or so for J.B. and me to write the 50 page pilot).

This is what it's like when you freelance, when you're not on the company teat. It's cold out here, but it's dynamic.

I'm just tired of making decisions like, Do I get that crown replaced or have the dog's lump biopsied?

Ever have this conversation with your significant other, driving your sick cat to the vet? "Listen, if it's over 250 dollars, it's sayonara, Sox."

Another meeting: There's a college football novel floating around out there—it's suddenly got some heat and I've got a crack at the adaptation.

I'd love to write a football movie. You know why? Because I'm tired of listening to basketball players saying how tough they are. Basketball players have no idea.

None.

You think Shaq would make an awesome lineman? Put Warren Sapp across from him for a set of downs and let's renew the conversation.

Football players, hockey players, boxers. These guys are tough.

I learned the hard way, on a Robin Williams football movie called, *The Best of Times.*

Ron Shelton wrote it, but an uptight Canadian named Roger Spottiswoode directed it, and it didn't turn out the way it might have.

Anyway, Ron knew I was struggling financially, so he got me on as a football player. They also hired some NFL guys, including Herman Edwards, who now coaches the Chiefs.

There was a linebacker from the Chicago Bears (I forget his name), and this guy was the most violent man I'd ever met. I went to lift weights with him at a local gym and he scared the crap out of everybody. He didn't lift weights, he attacked them. He started by benching 440 and screaming the whole time. I'd never seen anything like it. He cleared the weight room out in six reps.

Anyway, we shot the big football game over several nights out at Moorpark College in the middle of February. It was freezing. And it was supposed to rain at halftime, so they had these big fire hoses out there and they'd spray us between takes, and the field turned to brown soup and we were soaking wet, shivering, slathered in mud, and it was truly miserable.

Needless to say, the NFL guys all disappeared after the first night, with the exception of Herman Edwards, who knows what it means to honor a commitment.

So they brought in these gnarly former junior college guys who just couldn't let the dream die. I had one across from me on the line (I played a tight end, at six foot five and all of 215), and this guy was a mean sonofabitch.

He immediately started tearing me up. A forearm shiv up under my face mask. A bell ringer to the helmet. A head butt to my gut. He knew all the tricks.

He'd never been on a movie before, and he'd keep whaling away after they yelled "Cut!"—which can get really annoying. Plus he had no idea that we were usually so far off-camera that we could've been making like Tiny Tim vs. Pee Wee Herman and no one would've known.

I was in for a long six nights. After a particularly gruesome down, I went back to the huddle, whimpering, "This shithead's fuckin' killin' me." The guy playing tackle alongside me was a real-deal player—just a couple years out of the Pac 10.

A strong quiet type. "Yeah, I noticed. Lemme talk to him."

The A.D. called the roll, Kurt Russell yelled "On three—break," and we trotted back to the line of scrimmage, where my buddy said to my nemesis across from us,

"Hey, pal, take it easy. It's only a movie."

"Fuck you," the guy answered.

Wrong answer.

The ball was snapped, I braced myself for another onslaught when there was a blur of a chop block flying in front of me, and I suddenly heard this weird crackling snap and this piercing, strangled scream.

It took a good 20 minutes for them to load the guy onto a gurney and slog him off the field.

"Thanks," I said to the former Bruin.

"Don't mention it," he answered.

16

BEWARE OF EMAILS

"Umhlanga Rocks, South Africa, 1987. Apartheid. Three black kids watching from a distance. Dozens of white surfers out there, carving up one of the world's great breaks. The black kids? Not allowed in the ocean. Not allowed in the fucking ocean! Can you imagine that!?"

George is spitting these words out at me over coffee at a local joint.

"There's a movie here," he says. "My pal Mark and I want you to help us write it."

I'm skeptical. "It's so . . . foreign. Plus it's period."

"What, period?! It was 14 years ago!"

"All I'm sayin' is, man . . . that's a hard sell."

God, I'm turning into a studio executive. I hate when I get like this. So many reasons to say no. As a writer, as a teller of stories, the moment you start anticipating them, you're done. And they (the bastards) have won. They don't even have to do the dirty work—so many writers willing to emasculate themselves.

Anyway, I look for the bright side. George is one of the good guys. He cares. He's hard-core. Plus he's brilliant—incredibly well-read, plus well-versed in movie history.

I tell him that *Bend It Like Beckham* is the greatest weapon they've got in getting this story to the big screen.

Movie about a bunch of wacky Sikhs in London—working in restaurants and playing girls' soccer. Thirty-two million dollars U.S. Forty-four million international. Huge for an indie feature. Who'd a thunk it?

I tell George I'm interested. We'll meet again next week.

Why do I bring up this story? First, to chronicle the never-ending search for sports movie ideas, but second, to illustrate a point:

Never say anything nasty in an email.

It *will* come back to haunt you.

It *will* wildly exaggerate the degree of your anger.

It *will* always be there. Subject to prying eyes, company spies, and federal subpoena.

When I first signed on for my columnist gig at ESPN.com, the editors of Page 2 warned me about posting my email address. You'll get spammed, you'll get flamed. You'll get all sorts of bizarre proposals and weird advice.

No one will send you money.

Really? The rest I can deal with. But no money, that's a damn shame.

Through some unbelievable fluke of the ethernet, I've actually had more than a hundred thousand human beings read my column rant in a single 12-hour period.

So here's what I say in this week's column: Instead of wasting any more of your (and your bosses') time, if each of my weekly readers would just send me twenty dollars I'd have TWO MILLION DOLLARS—enough to make *94 Feet* and shut me up for good. (Plus it'd be the first movie to have 100,000 executive producers, which I think would be a record. I think.)

But there are laws against this kind of behavior.

Soupy Sales tried this once years ago on his afternoon kids' show (advising his young audience to find their dads' wallets and pull out those green pieces of paper with pictures of guys with beards and send them to him and they'd all get a postcard from Puerto Rico). It got him kicked off the air for a while.

You know why it's illegal?

Because the bankers and lawyers can't take their cut.

God forbid that people are allowed to invest in other people's wildly irrational schemes without the suits skimming all the cream!

Where was I? Oh yeah, nasty emails . . .

So a few weeks ago, I shot my mouth off about the politicians (Republicans *and* Democrats) so callously extending our soldiers' tours in Iraq. Indefinitely if need be. And how at least in Vietnam the guys knew how many more days they had to survive without getting their legs blown off.

Fairly benign stuff, right?

Wrong.

I get an email. From George of all people.

"Hey, dude, the last thing we need is more Hollywood pseudopolitical bullshit! Etc., etc."

It stings. The word "pseudo" means "fake, counterfeit" (harsh words from one writer to another). God, George, I thought we were friends.

So I break my first rule—Never say anything nasty in an email. I send him back two words, "Tough shit."

Big mistake.

Look, until now, every time I've gotten flamed in my column, I've written back, "Hey, sorry it doesn't work for you. Just trying to pay the rent. Best, Rob."

It works. People who flame other people generally don't like themselves. It's best to kill 'em with kindness. It throws 'em off guard. "Why is this guy being so nice to me?"

But George I knew. Or I thought I knew.

I first met him on the killing fields of AYSO. And for those of you who haven't been introduced to the Amerikan Youth Soccer Organization, be warned—it's run by parents who never got elected for student council (and not for lack of trying).

These people would gladly trade in their shorts and tennis shoes for brown shirts and goose-stepping jodhpurs.

But that ain't George.

I spot him at my 12-year-old's game. He looks like some dangerous lowlife sitting there in his lawn chair—unshaven, earrings, bandanna around his head. Leather jacket. Ripped jeans. Yelling at his kid, yelling at the refs.

This is how I like my male friends. Takes the pressure off. My wife gets on me about my personal grooming habits I can say, "Hey, look at George."

"George has a job," she answers.

"Barely," I reply. (He teaches screenwriting at AFI—the American Film Institute—once a week.) "And what about Peter? I saw him in Starbucks the other day, he was wearing his pajama top and gym shorts."

"Sounds hip," she says.

"Short gym shorts. With white socks and Birkenstocks."

"Well, that's Peter," she says.

"He doesn't have a job either," I say.

"So now you want to be like Peter?"

I hesitate, grope for a response. "We still clean up nice," I say lamely.

She stares at me. "Do you?"

Sometimes you can't win. You settle. Everybody settles. Whaddaya gonna do, kill yourself?

It's the quarter break at the soccer game. My son Cole runs up to me. "Did you bring my water?"

"No, I told you to bring your own water."

"Dad, jeez, Dad! I'm thirsty. I'm really really thirsty. Give me some money and I'll get a Gatorade!"

I consider. His face is flushed. He has been playing hard. I reach for my wallet, thwarted again. Another vain attempt to teach my child a critical life lesson—bring your own water.

Cole grabs the money and sprints off.

I watch and listen in as George's long-haired kid, Tristan, walks up.

"God, T, what's the deal!?" yells George. "You're so complacent out there. I'm ready to go home and grab a kitchen knife and commit hari-kari and you're gonna be blasé like that?! Gimme a break here!"

Tristan blows him off. Doesn't say a word. Swigs from his water bottle and returns to the field. How is it that kids are so much better at handling their parents than vice versa?

On top of which, they're brilliant negotiators.

I heard Herb Cohen interviewed on NPR about his book *Negotiate This!*

He was explaining why kids are the best negotiators out there. First, they ask for the impossible, raising the bar so high that any compromise works in their interest. Then they're absolutely relentless—don't know the meaning of no. They'll beg, cajole, seduce, threaten, and beg and beg and finally construct the most fantastic trade-offs—"I'll like do all your laundry for two weeks if you just buy me that Xbox."

Anyway, George has caught my attention.

I see some three-year-old wander by and I overhear George whisper, "Hey, kid, you got a couple bucks?"

"Huh?" The kid's totally bewildered.

"Gimme a couple bucks," says George. "Got a fiver?"

"No," says the boy, taking a step back.

"Oh, too bad—hey, where's your mom, your mom around here?"

The kid points and George looks. "Hmmm, nice-lookin' lady. I bet she's got five bucks. Go ask her, all right?"

The kid runs for his mother, who immediately starts shooting George alarmed looks and moves her chair down a ways. I figure it's time to introduce myself. George is my kind of guy. Maybe he'll help finance my movie—shaking down toddlers.

So we strike up a friendship. Share story ideas.

And just a few weeks later—he's flaming me in an email:

"Fuck you! Didn't you ever hear of a soldier's right to bitch!? Didn't you ever . . . " And on and on. Wow. This guy's a lunatic. Thank God I'm finding out early in the friendship.

But I'm bewildered—something doesn't feel right. I search through the in-box, the out-box, sent items, deleted items, saved mail. I find the answer in my address book. Aha.

There are two Georges. George number one is a true lunatic. The other George, my George, he's just out of his mind like the rest of us.

Thank God I caught it when I did.

Beware of emails.

They're inherent mischief makers. They send themselves to the wrong people. They confuse identities. They carry viruses. Their viruses carry viruses.

What was meant to be witty is read as virulent. What was meant as reasoned disagreement is misinterpreted as the demented workings of a sick, twisted mind.

Everyone's got an email story.

The young woman in England who sent her fiancé a sexy (and detailed) email about their upcoming night together. That thing traveled around the world in about six hours. With her name on it!

As my first-grader would say, "She was dweadfully embawwassed."

Anyway, it's a great relief. What an idiot I was to think that my friend George would go off on me like that.

And for the lunatic George who's floating around out there somewhere in the ethernet—you know I love you, right? And your mother loves you too.

So I'm back on board with *Zulu Wave*.

And I go to bed at peace. With an image I can't get out of my head. A group of black South African teenagers, hanging at the beach. Waiting till dark, when the whites clear out. Then dashing across the sand with their one battered surfboard. Taking 15-minute turns, an old alarm clock keeping it fair.

Surfing at night.

The ocean black. The kids on the shore staring hard—it's impossible to see. Then a wave breaks and the foam catches the white of the moon. And there he is, Kwezi Sisulu, shooting out of the curl.

And they cheer over the rumble of the surf at their feet.

And they cheer over the rumble of the war at their backs.

They're taking back the beach.

Surfing at night.

17

OVERNIGHT SUCCESS

My 12-year-old son Cole and I go to a UCLA game in Pauley Pavilion. Down on the floor before it starts, we pick up a free poster, then stand in line at the edge of the bleachers. The venerable old coach John Wooden sits in the second row signing autographs. Cole barely knows who he is. But I do. Because in 1969, we lost the final game of a Christmas tournament to UCLA on this very floor. By one point. The bitterness still bites. We inch our way forward and mumble graciously as he scrawls his name across the poster. Then I remind him of that game from years ago. "I remember it well," he says with a twinkle in his eye.

The first time I pitched a movie idea in Hollywood, I didn't have a clue what I was doing. I got the job. Sweet. I'd been in L.A. six months and I was two for two. I was still in the throes of development hell on my first gig when my William Morris agent called about an open assignment with a small independent production company.

"But I'm stuck in rewrites," I told her.

"Always line up your next job before you finish your latest," she responded (something I'd also heard from Walter Hill, coming off *The Warriors*).

She told me Aurora Productions was a small independent. They had a script called *Library Cop*. They liked the idea but hated the script.

They wanted a page one rewrite. If I came up with the right scenario, they'd hire me.

"What's the idea?" I asked over the phone.

"A young man wants desperately to be a cop, but can't pass the test, so he becomes a library cop."

"That's it?"

"That's it."

Wow. Is that the worst idea you ever heard for a movie or what?

But they were paying real dollars. Thirty thousand real dollars, as a matter of fact.

Library Cop—God, what a great idea for a movie!

I spent a few days working up a scenario, that figure echoing in my brain—30 grand, 30K, 30 thousand buckaroos.

So, here's the (too long) log line (which is essentially supposed to be a one-line *TV Guide* description) I came up with: Squeaky clean guy from the Midwest moves to L.A., fails the police academy exam, ends up in the downtown library, where he falls for a kinky older librarian and uncovers an elaborate cold war spy ring which is using the library to pass along secret documents.

Do you see now how money corrupts? I'm not talking values here, I'm talking brain cells.

Here's something that makes it easier: You come up with a crappy comedy idea—first thing you say is, "Imagine Adam Sandler playin' a guy who always wanted to be a cop but . . ."

See? It's called polishing a turd.

So I went into Aurora to pitch this thing to a woman VP and her gay French development guy. How did I know he was gay? He told me.

"You're so tall," the guy said. He turned to his boss. "Isn't he tall?"

"He's very tall," she answered.

Jesus, I thought, I am a long way from New Jersey.

I sat down and started pitching the story. They were with me for the first five minutes. They especially liked the kinky older librarian angle.

Although the French guy said (sounding suspiciously like Inspector Clouseau), "But of course she must be young too."

"You mean a young older librarian?" I asked.

"Exactly," he said.

"You mean younger in spirit, right, Jean?" his boss said.

"Is zat what I mean?" he asked, and they both laughed.

I sat there thinking, Is this guy even French?

But I plowed on—laying out, beat by beat, this overly elaborate spy plot, how they used certain books on certain days of the week, and if a book had been turned upside down it meant one thing, and if it had . . . blah blah blah, yadda yadda yadda, who gives a rat's ass? Certainly not these two, who started whispering about where they were going for lunch until finally, a good 30 minutes later, I was done.

"Okay, we like it," she said. "When can you start?"

"Tomorrow," I said. And I walked out thinking, Oh, man, did I just ace that or what?

Little did I know that they were both about to be fired from Aurora Productions and this would be their last project, and that *Library Cop* had as much chance of getting made as *Howard the Duck*.

Wait a minute, that did get made.

Well, then, as *Baseketball*. Whoops, that got made too. So did *Ishtar* and *Waterworld* and *Gigli* and *The Adventures of Pluto Nash*.

The Aurora people probably called the William Morris Agency and said, "Send us your tallest writer."

But I went away thinking, I'm gonna make this script work. I'm still thinking like that. You've got to. Which brings me to my current dilemma. In this town, everything takes forever.

So where am I at?

An email arrives from Mike Tollin. I recently made mention in my column of *Coach Carter*, a Samuel Jackson hoops movie that's shooting right now. How they might need some help with the basketball scenes. Tollin's one of the producers (a guy I've played ball with). He writes that they're more than happy with their basketball coordinator, Mark Ellis. Whoops. My bad. So props to Mark Ellis for making it work.

But Mike digs the column, and I'm glad to have reignited a dialogue there. Tollin/Robbins are *the* producers of sports projects in Hollywood. And a director I'm chasing, Paul Johansson, is acting in one of their TV shows, *One Tree Hill*.

I'm waiting to hear. Johansson's movie for Showtime, *The Incredible Mrs. Ritchie*, just received five Daytime Emmy nominations. So I may be waiting for a while.

I had a good meeting with Alex Gartner over at MGM about my interracial romantic comedy, *Take Me to the River*. Gartner is one of the producers on the *Barbershop* movies. I'm waiting to hear.

I've got *94 Feet* out to a few people. I'm waiting to hear.

But in Hollywood, you can't sit still. It's swim or die. I meet up with Todd Ramasar down at Staples during the Pac 10 Tournament. Todd's a former UCLA baller who's turned himself into a sports agent. He represents Baron Davis, which is not a bad place to start.

Ramasar wants to do it right. Keep it clean. Represent stand-up guys.

It's not easy. Guys still in high school and college with NBA potential will literally tell agents they need some cash up front. Not every guy. Some. Enough to turn the landscape into a minefield.

Years ago, I ran into an agent at the NBA Rookie Camp in Chicago. He'd gotten busted for slipping Marcus Camby some cash back at UMass. He was pissed because Camby only admitted to taking 2,000 dollars. "It was 20 times that!" the guy told me, as if that somehow exonerated him.

Anyway, the Hornets are coming into L.A. in a couple weeks, and I've been promised a meeting with Baron (who's turning himself into a real movie producer).

In the meantime, Spike Lee's developing a project with LeBron James. That's cool. You don't write a script overnight, so they've gotta be a year away. I'll reach out to Spike with *94 Feet*. "Hey, this one's ready to go. Wanna take a look?"

Unlike a lot of people who take an inordinate delight in dissing Spike, I think he's made some excellent (and important) movies.

Last week, in my column, I mentioned that if 100,000 readers each sent me 20 bucks, I'd have TWO MILLION DOLLARS! Enough to make *94 Feet*.

So far I've got promissory notes from eleven takers! That's the promise of $220. Only $1,999,780 to go!

Remember, in real life, this type of solicitation is highly illegal, but if we actually attain this goal, maybe we can ask the Bush White House for a special exemption.

Better yet, here's another idea. Did you catch the uproar caused by Maryland fans showing up against Duke wearing FUCK DUKE T-shirts?

The whole country was shocked. Shocked! So, for all you anti-Duke fans out there, for only $14.99 you can buy your very own FUKE DUCK T-shirt, thereby enabling you to stay warmly ensconced in the righteousness of family-value propriety and not get kicked out of school.

Or we can make it DUCK FUKE if you'd prefer.

T-shirts aside, for the moment I'm gonna concentrate on these two pitches:

One's a college football story based on a very funny (and timely) novel. The producer's pledged me to secrecy, so that's all I can say.

The other is the story of a Hollywood sports adviser. Actually, two Hollywood sports advisers, based loosely on my friend Kevin (K.B.) Benton and me. You know, do that black-white bickering-buddy thing.

You see, K.B. and I have worked together on several movies. *White Men, Blue Chips, Celtic Pride, The Sixth Man*. We've been through some wars.

K.B.'s an actor. That's his first love. Me, I wanna write and direct.

That's the premise of the movie. These guys land what to most sports fanatics would be a dream job—Hollywood sports adviser. But is that enough for these two knuckleheads? No way.

Above the line, that's where they belong. That's the hallowed ground where stars and directors and producers reside. Everyone else is below the line.

(Except for writers. They *are* the line. The thin line between boffo and stinko.)

For this movie, I've got a hundred stories in the bank. Crazed directors. Egomaniacal stars. Sexy cheerleaders. Obsessed fans. Wayward players. More than I can count.

So here's a little trip down memory lane. My reactions to events from a few of the many movies I've worked on:

The Sixth Man—"And why exactly did you get arrested? Selling calling card numbers? And why should we come bail you out?"

Eddie—"Wait a minute, Whoopi started sleeping with Frank Langella and now they're doing what to the script!?"

Celtic Pride—"Fellas, please, no guns in the locker room. Leave your guns in the car, all right, please?"

Blue Chips—(and this one was more serious, involving a female crew member) "He did what to you? Do you want to press charges? You don't want to press charges? Then why are you telling me this?" (Three nights later I see this same young woman stumbling out of an elevator with the player in question. Drunk, laughing, arms around each other. Think about it. We almost called the authorities.)

White Men Can't Jump—Venice Beach, about to roll cameras. I look around, one of our players is in handcuffs, being grilled by two cops. I hustle over, "Look, okay, so you think he assaulted somebody, I got that, but if you could just spare him for like 20 minutes—we need him in this next shot."

The player was a white guy, by the way.

Then there's the marijuana. Ah, yes. Even though alcohol's the drug that accounts for 40 percent of all violent crime in this country, it's pot they'll bust you for.

"Gentlemen, please don't get arrested for smoking marijuana. Don't smoke it on the set, don't smoke it in the locker room, don't smoke it in your car. Marijuana reeks. It sticks to your clothes. It reddens your eyes and fries your brain. It is illegal. It will land you in jail."

On *The Sixth Man*, up in Vancouver, Canada. Another speech. "We're all gonna be driving down to Seattle for six days of shooting at U Dub. Do not try to bring drugs across the border. Repeat, do not have drugs on your person as we cross the border. If you're black, chances are you will get searched."

There was a nasty-ass production manager on that movie. Didn't even try to make nice with us. She hung out with J.G., the line producer, and there was no love lost between them and the basketball people.

The next morning we're driving south in our rental. K.B. and me and our trusty African-Canadian assistant, Phil.

Caravanning toward the border, the line producer's car in the lead. Phil says, "Did you guys notice those players around J.G.'s Town Car, eh? And the duct tape?"

"What duct tape?"

"I don't know. I thought I saw one of them under J.G.'s car, eh?"

"You're kidding me."

"Afraid not."

"You don't think . . ."

"They wouldn't . . ."

We were about six miles from the border. The producers' car was about nine cars ahead of us. I floored the rental as K.B. grabbed the cell phone.

So what happened? Sorry, you're gonna have to wait and pay your eight bucks like everybody else.

But that, when you're prepping a pitch, that's where you start. Good material. Rich stuff. Stuff like *M*A*S*H*, *48 Hrs.*, *Jerry Maguire*, *Get Shorty*.

That's why I had such a hard time writing *Library Cop*. 'Cause there was nothing there. Nothing. Nada. Or *rien*, as the development guy would've said.

That job ended as absurdly as it began. By the time I had a draft to turn in, both the woman VP and the French guy had been given notice.

We had our final meeting at her house above the Sunset Strip. The French guy was in the dumps. The VP was even worse off. Besides her getting fired, her husband had left her, plus she was recovering from surgery.

She looked at me with pained eyes as I handed her the script. Putting on a brave face.

"In the last six weeks, I lost my job, my husband, and my uterus," she said.

"Is there anything I can do?" I asked stupidly.

(Like what, you moron, look under the couch for her uterus?)

"Just go away," she said.

So I did.

18

DUE DILIGENCE

To succeed in Hollywood you have to be resourceful, relentless, reasonably intelligent, and extremely lucky.

For those people who've been emailing me for advice, at best you're two out of four.

What'd they say about the Polish actress? She slept with the writer.

You're asking me how it's done? I don't know. (Am I whining again?) You flail away. You try this, you try that. You write like a maniac and see what happens.

If you're too calculating when you start, your stuff's gonna be sterile.

Otherwise, become an agent. Become a producer. Legitimate, necessary roles to be filled, plus you get to eat a lot of lunch and get lied to on the phone all day.

Besides, who knows what'll stick? (Might as well be somebody else's sweat-drenched script.)

Look, this is a town where Jason Alexander's about to star in a sitcom based on the life of Tony Kornheiser. Yeah, that Tony Kornheiser. ESPN? *Pardon the Interruption*? Don't get me wrong, I respect the man—he's a real pro. But you know how with some people, you say to yourself, Okay, I've already got TMI on this guy, I don't need to know any more?

Personally, I don't want to see Tony Kornheiser in his boxers fixing breakfast for his wacky kids. (Actually, I'd prefer to see him in his boxers than Jason Alexander playing him in his boxers.)

But say they got Jessica Simpson to play his daughter, and we get to see her in his boxers—now we're getting somewhere.

(Oh, hey, I got it, I got it—how about if she has to climb up on the kitchen counter to kill a spider and instead of wearing boxers she's wearing a skimpy T-shirt and . . . ?)

Why am I so grumpy today?

I just wrote and cut this ugly rant about Dennis Miller, how the guy totally screws up *Monday Night Football* and gets rewarded with his own show on CNBC where he demonstrates zero respect for his guests and how his shtick would be better served on shock radio but how lost he'd be without his smirk, blah, blah, blah, but like I said, I cut it.

I'm not like that.

Life's too short for that kind of negativity.

Dennis, I'm sorry, man. You know I love you, and I'm only illing you like this because I know your show's gonna be canceled before I get invited on.

But I actually dig the emails. I do, I swear. All of them. I try to answer each and every one. But the ideas . . . I cannot handle the ideas.

Not that they're no good. It's just that I can't handle them.

Except the one about al Qaeda infiltrating the Harlem Globetrotters so then the FBI has to infiltrate the Washington Generals and they play the big game at the United Center and George Bush shows up because he's a longtime Globetrotters fan and all hell breaks loose.

Now that, that's an idea.

You gotta love all these writers out here, brimming over with creativity.

But here's the problem—it's a numbers game. Entertainment has become nationalized. And in a nation of 290 million people, it's a pretty slim sliver at the top of the pyramid.

I had these two friends from Iceland. They'd become a big hit out there in the North Atlantic. Certified stars. National Treasures, these two.

But how big is Iceland? Let's see, about the size of Hackensack, New Jersey.

And how long did these two Icelandians last in Hollywood? Long enough to melt.

Regional theater, that's where it's at. Garage bands. Poetry readings. All those places where you'll find ABSOLUTELY NO MONEY.

It's where the artists go.

These are the real American heroes.

They're doing. They're teaching. They're teaching gym. They're making art. They're playing Stanley Kowalski and high school volleyball and sax in the marching band.

They're not on television.

See, it's all about the dignity of the regular guy and the average gal, the ones out there plugging away, making this great country what it is today. Firm in the belief that, working together, they can put meat on the table, find love between the sheets, and still write a screenplay on the side.

All I can say is, go for it. Development people have no clue where the next great screenplay is gonna come from. So prey on their insecurities. And please, go visit www.breakingin.net if you need more info. Or type "screenwriting" into Google and you'll find all sorts of advice from people way more knowledgeable than me.

But know this—your first effort isn't going to be as good as you think it is.

(Yeah, you say, tell that to Matt and Ben about *Good Will Hunting* all the way to the Oscars. Yeah, and I'll tell you that the studio execs threw away half their screenplay because it had Will going to work for the CIA and saving the world from some fakakta nuclear bomb threat or something.)

Here's another reality check which might work in your favor. These days, Hollywood's more likely to buy your idea if it exists in another form.

An illustrated book, a cartoon character, an article from *Rolling Stone.*

Or *Boondocks*—my favorite comic strip. I'm sure there's a TV deal happening there.

And don't forget Tony Kornheiser, sports show personality, as sitcom dad.

A novel, a play, a song.

A column. Aha. A column.

You know what? For me, it's actually working.

I recently mentioned a producer (I'll call him J.C.) who contacted me weeks ago (cold email). He's the real deal—he's not too big, but he's solid. He's been a studio executive, he's got projects set up, great representation at Creative Artists Agency (CAA).

He feels like the partner I should've had in my early years in Hollywood.

We're working together on a college football pitch, but still in that due diligence stage regarding other projects.

What's due diligence, you ask (again)? Well, you know how before you marry the woman of your dreams, you make sure she doesn't have some psychotic ex-husband peeing around the perimeter? Or some stoned-out brother who's gonna camp out in the backyard like Randy Quaid until she breaks down and fixes him up a room in the basement? Or that she's not actually a man? That's due diligence.

When people are young, they sleep together before getting married to find out how good the sex is gonna be. Second time around, it's to find out how bad's the snoring. Due diligence.

Think Jerry Falwell's seen that new study showing that teenagers who pledge themselves to abstinence before marriage are nonetheless getting STDs at the same rate as all the immoral American teenagers?

Ah, America. And the Bill Clinton definition of sex.

These days, it's not the act that counts, it's the pledge.

You get to feel so righteous, and you still get to go out and hump like bunnies, just like Jessica and Nick (only they were actually married).

Damn the hypocrisy, full speed ahead.

(For those of you under 12, STDs are a lot like WMDs only you can find a whole bunch more of both in America than in Iraq.)

But suddenly for me, right now, it's working. People are paying attention.

I get an email from Rat Entertainment. It's John Cheng, head of development. He digs my rants in my weekly column.

I'm thrilled. Because Rat Entertainment is owned by Brett Ratner, who directed both *Rush Hour* movies, which my kids and I enjoy no end, plus he's done a zillion of the hottest rock videos and he's tight with Russell Simmons and he's tight with Shaq and the guys, and he's filled with exuberance and moxie, really in love with the movie-making process, and I want to do business with him.

I call Rat Entertainment, hoping for a quick meeting. John Cheng calls me back. He's on location in Edmonton, Canada, won't be free to meet for five weeks. Okay, that's cool. That's Hollywood. Everything takes forever.

At least I'm making some headway here.

Plus there's the book: *Hollywood Jock,* to be based partly on my column. My book agent, Matthew Guma, submitted the first eleven columns to Hyperion Books a few weeks ago. (They're owned by Disney, just like ESPN—they get first look.)

Waiting to hear. (Oh, Gretchen, we're waiting.)

I want Guma to put some heat on Hyperion but I'm afraid to call him because he's a Carolina Tarheel and he probably committed hari-kari over the weekend since the Heels got trounced.

But it's obvious we're gonna sell a lot of copies here. Look, you're mixing Hollywood and sports—that's a huge audience.

But Guma cautions me, "You gotta remember, guys don't read. Guys don't buy books."

"Then look," I reply. "Here's how we sell it. I'll go on *Oprah*: 'Ladies, do you wish your man would crack open a book once in a while instead of watching Tom Arnold play one-on-one with Gary Coleman on *The Best Damn Sports Show Question Mark?*'"

"It's a natural. Buy him *Hollywood Jock*—the book for guys who don't read books."

Plus I'm obviously willing to shamelessly self-promote and whore myself (absolute requisites in this cluttered market).

Plus, plus, it's following on the heels of mad screenwriter Joe Eszterhas's *Hollywood Animal* and that hit the best-seller lists big time.

Only he's got scenes of naked women wearing mink coats delivering script notes from Robert Evans tucked neatly inside their um, their . . . you know. While I've got scenes of my wife calling me during a casting session on *Eddie* to get my ass home immediately because our four-year-old's got diarrhea again, plus the thingy says she's ovulating and if we're gonna have that second kid . . .

Actually, that's not bad—get, say, Jennifer Lopez to play my wife and next thing you know, our hero's dashing home for a quick, sweaty roll in the hay, and suddenly . . .

It's a column and a book and a movie!

Actually, the movie idea's been in the works for years. (It's called separation of rights; we don't want anyone getting sticky fingers here.)

I'll leave that to Joe Eszterhas.

Hey, Brett, you directing this thing or what?

All right, time to take a deep breath.

I'm sitting in a booth at Art's Deli, talking through the project with my new producer pal, J.C., who also thinks Ratner's a great choice to direct.

"And you know who'd be perfect to play your character?" he says.

"Who?" I ask (thinking to myself, George Clooney, Brad Pitt, Benicio Del Toro . . .).

"Ben Stiller," he says.

"Ben Stiller?"

"Ben Stiller."

And I say to myself, Hey, why not just make it Jason Alexander and call it a day?

19

THE BEST DAMN SPORTS SHOW QUESTION MARK

The Twentieth Century-Fox lot is on Pico Boulevard, West L.A. I've got a meeting. It's the wrong century, but I show up anyway, pulling in, joining a short line of Mercedes and Lexuses.

In Los Angeles, you are your car. I'm in the wrong town.

I got my car attitude from my old man back in New Jersey: "Buy a cheap, reliable used car, drive it into the ground [we're talking 200,000 miles here], then buy another one."

I can't get it out of my system. I wanna send my kids to college. So I usually park around the corner. Or borrow my wife's new minivan—at least it's not all dinged up and droopy-assed like my old Toyota.

Security on most studio lots is pretty much the same. A guard booth, a crossing arm. You pull up, give your name, and they check if there's a drive-on for you.

I know it'll be there, because I called Robyn an hour ago to remind her.

Half the time you get to the gate and the assistant forgot to call it down and you gotta pull over while security calls upstairs and it's a big pain in the ass (particularly if you're sitting in your wife's minivan as the Beemers and Jags cruise by).

So you learn to call ahead. You learn the assistants' names. I've even seen screenwriters bring flowers, but that's going too far in my book.

Even with my wife. I explain to her that flowers seem like such an obvious bribe.

"What are you, turning into Larry David?" she says. "Flowers aren't a bribe. A trip to Hawaii, that's a bribe."

Or how about a four-million-dollar diamond ring?

I heard a comedian say, "Nice move, Kobe, now every time she looks at it, she'll be reminded of how you cheated on her."

Maybe. But my bet is that she'll be reminded of how rich she is.

The guard slaps a pass on my windshield, hands me a map, and tells me to park in Lot B—it's next to one of the newer buildings.

Aha—this is where they shoot *The Best Damn Sports Show Question Mark.*

My actual meeting's across the lot with the guy who runs Fox Television Pictures, but I ran into John Salley a few weeks ago at that NBA All-Star party and I wouldn't mind bumping into him again. (Salley's on *Best Damn*, for those of you who've killed your televisions.)

Guys like Salley move in interesting circles. Circles of celebrity. And quasi-celebrity. Circles that drift and shift and overlap. Music, TV, sports, movies.

These people show up at the hippest clubs, at Laker games, in snazzy restaurants, where they get to have fun saying hello, catching up, talking some trash, knowing that all eyes are upon them.

And where they get introduced to each other, creating even more circles.

It's easy for them to become dismissive of noncelebrities.

It's like high school.

The "in" crowd.

But in high school, anybody can say, "Hey, so what? I'm too cool for you anyhow. I'm starting my own club. I've got my own friends."

Yeah, but do you have your own TV show?

I see a guy coming out the building. He's got an ID card hanging from a string. He's got the look and jaunt of a smart jock. I call out to him. "Hey, is Salley here yet?"

"Nah, he gets in around one o'clock."

"This entrance?" I ask.

That gets his antenna up. "Who are you?"

"I worked with him on *Eddie*. Whoopi Goldberg coaches the Knicks?"

"Was that movie your fault?"

"No, I swear. I just did the basketball. With Kurt Rambis. Let's blame him."

The guy laughs. "Salley does the valet. They've got a one o'clock production meeting every day."

"Thanks," I say. "Maybe I'll run into him. I've got a meeting across the lot."

I walk off thinking, What're you gonna do, cut short a meeting with the head of Fox Television Pictures for a chance to bump into John Salley in a parking lot?

I got my first real studio screenwriting deal on this same Fox lot more than 20 years ago. And the guy I'm about to see, David Madden, is the guy who hired me.

A movie had just come out, *Cannonball Run*, and it was a huge success.

Fox wanted to do a high-concept takeoff—*Cannonball Run Around the World*. And they were calling in screenwriters to see whom they'd hire.

This is a common way for movies to get written (and almost always never made). They put the word out to the agencies that they've got an open assignment, and the agents send them appropriate writers.

Who decides who's an appropriate writer? The Appropriate Writer Fairy.

You meet the executive, he pitches you the idea, the two of you riff on it for a while, then you go away for a few days and figure out your approach, then you go back and pitch it back to him.

So, in our first meeting, Madden said, "*Cannonball Run Around the World*."

I thought about it for a minute, then said, "There are cars in *Cannonball Run*. They're racing across America in cars."

"Yeah . . . ?"

"So for an around-the-world car race, what happens when they reach the beach?"

"That's why you're here. To answer that question."

And they do this with several writers. A bunch of writers. And the writers go away, then all come back with their takes, and if you're paranoid that they're gonna steal your best ideas and hire someone else, this isn't the game for you.

After a while you learn not to worry about it because none of these movies ever get made anyway.

So I went away and came back the next week.

"Forget the cars," I said.

"Good," said Madden.

"Make the characters explorers. Old school, you know, Sir Edmund Hillary types, and new school—young brash Turks."

"Yeah, I'm with you. And . . . ?"

"Start with a black-tie dinner at the Explorers Club in Manhattan. Huge uproar over who're the best explorers. They devise a contest. An around-the-world test of endurance, smarts, adaptability. Jungle, mountain, river, primitive cultures, wild animals . . ." (I was pitching this in 1981, mind you.)

"Uh-huh, uh-huh."

"Start and finish in New York City. I want to call it 360."

"I like it," said Madden. "I'll call your agent." (It wasn't exactly this clean, but it was close.)

Walking off the Fox lot that day . . . God, that was a good day.

That was an "Oh, the places you'll go! There is fun to be done! There are points to be scored. There are games to be won" Dr. Seuss kind of day.

Six months (and about 40 thousand dollars) later I turned in a pretty cool script.

Madden liked it. But he was leaving Fox for Paramount. And just like that, 360 was dead. Aw, man.

This happens all the time. Executives will develop scripts, then change jobs, and the first thing the new executives do is throw away all the old scripts.

But that was then and this is now.

And David Madden's one of the good guys—intelligent, straightforward.

I'm always slamming Hollywood—but there are actually a lot of good people in this town. You'll probably find more of them in television than in movies. TV people have to work so damn hard, they don't have enough time to be overly neurotic.

I walk into Madden's outer office in Building 41. It's filled with scripts, floor to ceiling. Endless screenplays, stacked there like so many orphaned, handicapped children. Second acts that don't work. Faulty structures. Cardboard characters.

Stacked there like corpses in a morgue.

But wait a minute. This is television. The insatiable medium. Some of these scripts are actually gonna get made.

Madden runs a division called Fox Television Pictures. You'll see their movies on Fox and FX, but also on CBS and Showtime and the WB and all over the place.

At first, it doesn't make sense. Why would one company make movies for a rival studio or network? The answer is simple. Because everybody still makes money.

Madden and I get caught up. He's spent years in features. He tells a story of how it took eleven years to get *Runaway Bride* made. The movie made money, it was a success. But eleven years—that takes a toll.

Movies made for television, even though the money and prestige are not nearly as good, at least there's an immediacy to the process. The networks are either gonna make it or they're not, and you find out fast.

Madden tells me how it works.

His division comes up with a diverse slate of movie ideas and/or actual scripts, then they shop them around to the different networks. If a network bites, it puts up a majority of the production costs, Fox

Pictures puts up the rest (totaling around five mil per movie), and they share the revenue.

He tells me they made one of my favorite cable movies—*Big Shot: Confessions of a Campus Bookie*. This bodes well. *Big Shot*, directed by Ernest Dickerson, hit the perfect tone—greed, ego, and lunacy, spiraling comedically out of control, but with a real edge to it.

Tough stuff to pull off.

Especially with a short, frenetic schedule. Most TV movies are shot in 18 to 22 days. It's a scramble, particularly if you've got lots of characters and locations. As Madden put it, "For every car crash, you'd better follow with five pages of two guys talking in the hospital."

Whereas feature films, studio films, have at least twice as many shooting days, and that number can easily climb into the sixties and beyond.

I did the basketball sequences on both the *Father of the Bride* movies with Steve Martin, and that pace felt almost leisurely.

Charles Shyer directed; his wife (at the time), Nancy Meyers, produced; and they both wrote. They were known as the Shmyers.

I'd sit on the set and watch Nancy stare into the video monitors after a take and whine, "Char-ar-les, don't you think Steve's collar looks too pointy? Look at it. It's too pointy."

"His collar's fine, Nancy," Charles would say.

"It's too pointy. I'm calling wardrobe."

And the whole thing would come to a halt, 110 people standing around as wardrobe fixed a pointy collar.

I shouldn't knock them too much. They actually made some good movies together. And if there's one thing I'll pound on again and again, it's this: Making a good movie is extremely difficult—there are simply too many things that can (and will) go wrong.

As for the whining, directors have different styles. I heard Garry Marshall once describe his approach to directing: "I beg. I beg my actors. I beg the D.P. I beg the studio. I beg the caterer."

Nancy Meyers finally broke off on her own and wrote and directed *Something's Gotta Give*, which is a decent movie and a tidy financial hit as well.

Go figure. In Hollywood you can actually whine your way to the top (as long as you put in the work).

I talk to David Madden about several scripts. First my college hoops movie, *94 Feet*. Then a spec script that Spielberg's Amblin Entertainment almost bought—perfect for the Sci-Fi Channel. An anthropologist and her husband and child stumble across an ancient tribe, living underground in New Mexico. Then there's a Texan P.I. action piece Walter Hill paid me to write. Plus my interracial river-rafting romantic comedy.

Madden's interested.

I mention the stacks of scripts piled everywhere. The endless reading required of these people. "How about if I just send you a paragraph on each project? If you like it, I'll send along the script."

"Great. Let's make something work here."

"Yeah, let's do that."

We part with genuine affection. I walk back across the lot, buoyed. Not "Oh, the Places You'll Go" Dr. Seuss buoyed, but buoyed nonetheless.

It's just after one o'clock. I look for John Salley by the Lot B valet parkers.

A luxury sedan pulls up. It's Tom Arnold. He gets out and sees me staring.

I see his mind racing with—Do I know this guy? No. Does he look like a threat? Nah. Do I say hi? Yeah, why not? "Hi," he says.

"Hi," I say back.

A quick smile, then he hustles off, late for the production meeting.

I spot the smart jock guy I spoke with earlier. We introduce ourselves. Turns out, he's Tomm Looney, the voice of *The Best Damn Sports Show Period*.

He's also working sports talk radio.

I tell him what I'm up to. Tell him I'd like to come on his radio show. Tell him how the last time I was on the radio, it was in Atlanta and the two jocks were a couple of real assholes and it was like, "So who'd you bang? Who's a jerk?" for ten straight minutes till I couldn't take it anymore.

Real morons, these two—the worst of ambush interviewing. And what made it worse, I played along instead of hanging up.

Looney says, "Yeah, I know those guys. They're good friends of mine."

"Oh, great," I say, thinking there goes any chance of appearing on *Best Damn Sports Show*. But Looney laughs, and I wonder—Is he messing with me?

"Are you messing with me?" I ask.

"No," he says, winking.

20.

SENOR, I DON'T THEENK SO

Beverly Hills. I'm driving my beater down Wilshire Boulevard. I pass the Regent Hotel, where I've got a meeting with Baron Davis.

I spot the valet parker but decide to find a meter instead. I can just hear the guy as he considers my car: "Senor, I don't theenk so. . . ."

I hang a right on El Camino. Familiar turf—the William Morris Agency looms big and black at the far end of the street.

The black hole.

I climb from the car and start quickly pumping quarters into the meter. I can feel the hum of the building at my back—"My force field is sucking you in. You have no strength to resist. Soon you will be within my bowels and no one will ever hear from you again."

Actually, the agencies have changed. Used to be you could pick up an agent quite easily. They'd throw a few of your scripts against the wall. If something stuck, you were on your way; if not, it's the black hole for you, baby.

Now, the agents have gotten a whole lot pickier. Who can blame them? All the crappy scripts everybody has to wade through. From Santa Monica to Burbank, they're stacked up on the curbs, spilling out of elevators, clogging the hallways.

I somehow break the force field and head toward the Regent, carrying my snazzy faux-leather briefcase (Staples, $29.99), prized screenplay inside.

I run into Baron Davis and his young black agent, Todd Ramasar, in the lobby. The Hornets are in town to play the Lakers. We head up to Baron's room and get down to business.

Baron Davis is actually pumping some energy and money into several movie projects. Thoughtfully, systematically turning himself into a producer. He's also looking to make movies down in Louisiana.

I lay out *94 Feet* for him. "In a world of cookie-cutter sports movies, it's an original—the story of one college basketball game. Not only do we never leave the arena, the script's got two endings. The plan is to shoot them both. The first sports movie ever where either team might win and only the final shot will tell."

Baron and Todd are intrigued.

I tell them I'm not here looking for money; I'm looking for involvement. Baron says he'd be interested in playing one of the basketball players if the schedule works.

This is good news. This guy's an NBA All-Star. Maybe he'll play a producer's role as well—helping attract other quality players. Players young enough to play college guys. (I'm open to suggestions.)

I want to get Luke Walton. I think that'd be a hoot. And Dwyane Wade of the Miami Heat. T.J. Ford. Like that. Carmelo and LeBron, obviously, but that's a tall order.

Besides, there'd be something cool about doing it with some of the unsung guys. Give it more of a hard-core attitude. And it'd be sweet to attract a couple of the *And 1* streetballers. Like the little white kid known as "The Professor." Or "Hot Sauce," or maybe "Skip to My Lou."

Of course, we'd have to get all of that street junk out of them.

So, real players, playing real ball.

But for the coaches, we'll need actors. I'd like to see Michael Chiklis of *The Shield*. And Lou Gossett Jr. to play his wily old nemesis.

Baron digs this. It's fun to talk casting.

Then we get into arenas. He brings up Tulane University in New Orleans. It'd be perfect. Small enough to fill with extras, big enough to look right. Plus there are real financial incentives to shooting in Louisiana—tax breaks, looser overtime regulations. (Of course, as of this writing, nobody'd anticipated Louisiana's wild night in a dark alley with a couple of Cajun honeys named Katrina and Rita.)

Baron asks for the script. I pull it from my briefcase, hand it over, shake hands good-bye, and walk out.

The screenwriter's life.

How many times have you done this? How many meetings do you walk away from thinking, That's the one, that's the one right there?

You get knocked down, you get up again.

You get knocked down, you get up again.

Nam myoho renge kyo.

I came to Hollywood well-prepared. I started in the movie business in New York City as a production assistant. Then *The Warriors* came to town, and I learned firsthand what it's like to get knocked down. Literally.

Riverside Drive Park, midnight, July 1978.

Here's a story I started weeks ago and never finished telling. It's hot and sticky. I'm wearing a baseball uniform, black cap, my face painted purple and black. The sweat's running streaks down my neck.

I've got a 36-inch Louisville Slugger in my hand, face-to-face with Michael Beck—"Swan," leader of the Warriors.

Craig Baxley, the stunt coordinator, is breaking down the sequence.

"Look, it's got eight steps. First—Rob, you swing high; Swan, you block. Then, Swan, you take a full cut at his head; Rob, you jump back. So let me see that much at half speed."

We practice. Baxley adds two more moves. We practice, picking up the pace. "Easy!" Baxley barks. We slide back into slow motion. The director, Walter Hill, and his D.P., Andy Lazlo, come over and observe, quietly conversing about coverage, camera angles. Then they walk away.

Baxley, Swan, and I keep at it. Once we get it down and work it up to speed, the sequence is gonna end with Swan taking a full cut, catching me in my midsection, sending me flying.

I'm wearing rib pads under my baseball jersey. Not exactly full metal jacket protection, and I'm wondering how this is gonna all go down. Walter delighted in testing guys' toughness. Especially Ivy League basketball players. Hence my trepidation.

I'd witnessed a nasty bat fight in real life, and it wasn't pretty.

It was in 1975, three years out of college. I was living a shabby existence in a crappy Second Avenue apartment. This was my penance for two years of cavorting in the Rocky Mountains with a bunch of other free spirits who were convinced the world was about to end. (After a while you realize, the world never ends. It just gets crappier.)

"Hey, God?! We're over here! Remember us? The Grand Experiment?! Save us from ourselves!"

"Save us from Osama!"

"Save us from Mel Gibson!"

Anyway, I bounced into New York City from Gold Hill, Colorado, and shacked up with an old Princeton teammate, Dominic Michel. I still had a jones for basketball that I was unable to shake.

I was hooping anywhere I could find a game. Riverside Park, West Fourth Street, Central Park behind the Met. Rough-and-tumble. Street ball. But it wasn't enough. And I didn't have the chops to make the NBA.

So I called another old teammate, John Hummer, who'd been playing in the NBA for the Buffalo Braves. "I wanna go overseas," I told John.

"I'll call my agent," Hummer answered.

A week later, I got a call. Hummer's agent. He'd landed me a job, sight unseen, with a team in Finland. September through March. Three hundred fifty dollars a week, an apartment, and a car.

But it wasn't even Helsinki. It was some town above the Arctic Circle. Swear to God. I said I'd take it. The agent came on strong. "Once you get over there, you gotta stick it out. There's no running home."

"No way," I answered. "I need to play some ball."

So I spent the summer working out. It was weird. It was a time in my life when I was truly lost. My first true love had left me. I had a huge chunk of an ill-conceived novel sitting alongside my battered electric typewriter. I was broker than broke.

The Vietnam War was still raging, tainting every good moment.

I'd tried to do my part. We'd marched in Washington. Marched in New York. We'd harbored draft dodgers in a Princeton dormitory—scared, lonely guys trying to make their way to Canada. We boycotted classes. Shut down the university. (Oh, now that's effective—let's stop a war by playing hooky.)

Still, the bodies kept coming home, week after week—132 American dead, 104 American dead, 97 American dead. Every week. It was horrifying. More than 57,000 total. In a war that anybody with any brains knew we couldn't win.

As a jock, it was particularly tough. Being a jock back in the '60s, you might as well have been in the army, the way people looked at you. Subjugating yourself to the system with the same unquestioning loyalty of those poor guys dying in Vietnam.

But some jocks spoke out—Muhammad Ali, Dave Meggysey, Rosey Grier, Jack Scott, Harry Edwards. John Carlos and Tommie Smith at the '68 Olympics. Bill Russell. Jim Brown.

And the world listened.

But not the boys in power, LBJ and McNamara (good Democrats, mind you). They still plowed on, up to their ears in blind hubris. While the body count kept rising.

Walk into any high school gym and somewhere you'll find a dusty plaque on the wall, honoring the American boys who died in Vietnam.

I just recently saw the name of a guy whose brother lives on my street. Chad Charlesworth. Killed in Binh Dinh, 1970. I mentioned the plaque to his brother, Allen.

"Yeah, he was 22 . . ." That's all he could say. The pain still ran that deep.

During Vietnam, the argument was made, week after week, "If we quit this war now, then these guys will have died in vain."

Sound familiar?

So yeah, I was lost in New York City.

Maybe basketball would save me.

One afternoon, I was down at the cage on West Fourth Street and Sixth Avenue. Contrary to popular myth, the games down there have always been erratic.

One day you'd find yourself running with Bernard King. The next day, you'd have some fat-assed, drunk, belligerent has-been posting you up like some Chocolate Thunder wannabe.

The school of hard knocks. I'd do a solitary workout in the mornings. Shoot three hundred Js. Then go find some crazy game in the afternoons.

It was twilight. I'd had a decent run at West Fourth. I was walking through Washington Square Park, brown-bagging a quart of beer. The drummers were out. The cool-cat chess players. The hustlers, the girls in short shorts. The swirling packs of kids.

It was New York City at its finest. All those rich folks could have the Hamptons. We had Washington Square Park.

I grabbed a spot on a bench, checking out the scene. Enough cash in my pocket for a subway token and two slices of pizza from my favorite joint on Second Avenue.

Clouds of marijuana smoke drifted over the throbbing music, over the mélange of young bodies out looking for action.

Then that quick, the vibe changed.

It started with the jarring crash of a broken bottle. Then another. Everybody froze. Heads swiveling. Suddenly there were a good 30, 35 young teenagers, angry, shouting, stalking from trash can to trash can, yanking out empty quart bottles, grabbing the necks and smashing the bottoms off.

A turf war. Choose your weapons. Broken bottles.

The kids were mostly white, Italian descent, a couple black kids in the mix. Shouting at each other, working up the anger. The onlookers swiftly dissipated, melting off the cobbled walks onto the grass, under the trees, away from the fountain. While the angry pack of young thugs swirled into a tight, threatening mass, waving their broken bottles, amping up their war cries.

I sat tight on my bench. Watching intently as I finally spotted the object of their anger. Four young teenagers. Puerto Ricans. Standing on the path at the northeast corner of the park. Each held a baseball bat. They were calm. They were resolute. They watched silently as the growing pack of teenagers worked themselves into a frenzy, then started marching toward them.

The four Puerto Ricans stood motionless. Then they charged. Four kids with baseball bats—sprinting headlong into a pack of teenagers wielding broken bottles.

That quick, the huge pack broke and ran. Throwing their bottles away, hopping benches, sprinting for the trees. And I heard the *crack crack crack* of the bats hitting young skulls, crashing down on fragile collarbones.

And that fast, it was over.

Six, seven kids, sprawled out on the walkways. Moaning. Crying out. The blood spilling. The sound of sirens.

Three years later, this memory runs through my head as I test the weight of the Louisville Slugger in my right hand.

The scene is set. Swan and I face off under the bright white klieg lights. The A.D., David Sosna, calls for quiet, then calls, "Roll sound."

"Rolling."

And the two camera operators call out, "Speed."

"Speed."

Walter Hill takes a moment, then quietly says, "And . . . action."

Swan and I swing and duck through our routine. The last piece—I raise my bat high, and Swan takes a full swing, catching me in the solar plexus.

I fly backwards, land on my ass, and roll onto my side. Gasping for the wind that's been knocked clear from my lungs.

I vaguely hear Walter yell, "Cut."

Swan and the stunt coordinator hover over me. "You okay?"

Here's the deal—stuntmen never show they're hurt. Never. You end up with a splintered tibia sticking six inches out of your shin you say something like, "Aw shit, I hate when that happens."

Now for an entire generation of suburban white boys who grew up just yearning to show Mommy their latest boo-boo, this can be a jarring new reality.

But somehow, staying curled up in the grass in a fetal position on a movie set just doesn't cut it.

Jim Brown once said, "When you get brought down, always get up slow. Don't bounce up—that's stupid. Get up slow, so when you're really hurt, nobody's gonna know."

I get up slow.

Walter looks from Swan to me. "That's it? Let's do it again."

So we do it again. And again.

After the third take, I'm really hurting. Baxley comes over. "Think you can go again?"

My ribs are on fire. I can't get enough air into my lungs. "Oh, man, jeez, Craig . . ."

"You know you get a 50 dollar stunt adjustment for each pop you take?"

"Yeah?"

"Yeah."

"Fifty dollars?"

"Each take."

I bounce to my feet, waving my bat.

"Hey, Walter, that last one sucked, let's go again!"

Like I said, good preparation for the life of a Hollywood screenwriter.

GOD AND SPORTS AND WAR

Last week in my column I took a cheap shot at Mel Gibson. It had nothing to do with *The Passion of the Christ*, I swear to God. I haven't even seen it, so I'm not about to knock him for it.

Look, I'm a recovering Catholic myself, and it just so happens I'm writing this on Easter Sunday, so you know that today of all days, I'm not gonna be messing with the Big Guy upstairs.

But a reader called me on it: "I'm pretty open-minded but you've got to realize there are obviously more Christian sports fans out there than you think."

Hey, I realize. How could I not realize when these guys are reminding me every time I watch a game? "First of all, I want to thank God for making this victory possible."

Victory—what victory? The fact that some poor bastard missed a free throw and you guys won the game?

Look, the beauty of sports is that for every winner, there's a loser.

This winner-thanking-God stuff, I don't know—I just don't buy it. Who are the losers supposed to thank?

Sometimes I think the Greeks and Romans got it right. Multiple Gods.

Bloopis—the God of missed fly balls.

Embarractis—the God of the right-angle golf slice.

Steroidus—the God of strength.

Soclosis—the God of the one-point defeat.

Get my drift?

When it comes to God and sports, if you can't at least rely on the Big Guy's being totally neutral, then it's all Greek to me.

Not that you can't call on Him to help you do your best. That's legit.

"Please, God, give me the strength to kick this 52 yard field goal, even though our offensive line has turned into Swiss cheese and I've been blocked once already and my nose is still bleeding, and 65,000 people are gonna hate me if I miss, not to mention the fact that my girlfriend's not gonna give it up for at least the next three weeks."

God's probably up there wondering, Why do you people put yourselves through this sports stuff anyway? Didn't I make things tough enough down there?

When I went to email this reader an answer, I figured I'd explain that it had nothing to do with Jesus or the movie or the fact that Mel Gibson is one of the worst actors in the business (especially when he's trying to be funny) but just that if you've ever seen him on a talk show, you know he's an obvious lunatic.

An extremist to the nth degree.

The world needs far fewer of these people.

Plus I knew he ran conservative when it came to politics, and if there's one thing I can get worked up about (besides Shaq being allowed to bull his way right through the poor sap trying to defend him) it's social justice (or the lack thereof).

And how if the rich greedy bastards of this world don't start sharing the wealth, the rest of us are gonna get mad as hell and decide not to take it anymore.

(For more info please read: *Perfectly Legal: The Covert Campaign to Rig Our Tax System to Benefit the Super Rich—and Cheat Everybody Else*, by David Cay Johnston. Amazon.com, $18.17.)

Hey, sports fans, contemplate the incredible tax breaks for the mega-owners and super-rich athletes the next time you pay $80 for a lousy seat and $15 for parking and $4.50 for a soggy hot dog.

Not a message your average free-market Mel Gibson-type conservative appreciates. Plus I assumed that Mel was an obvious supporter of the war in Iraq, so in my email to my Christian critic, I put that in too. And I ended by adding, "Besides, what would Jesus do?"

But I hesitated before firing it off, as one of the preeminent rules of life came back to me—ASSUME NOTHING.

So I ran a Google—mel gibson iraq war.

And lo and behold, the very first item—"Gibson hits out at war in Iraq."

Jeez. Who knew?

Suddenly there's hope. Mel Gibson has given me hope, and it didn't cost me eight bucks. There are others in the worlds of sports and Hollywood who have seen this writing on the wall.

(And I'm sorry if I'm ruining your coffee break, but in this world where one of the best Super Bowls of all time can get lost in the furor over some idiot's right breast, the least we can do is snap out of the crap once in a while and pay attention to things that really matter.)

Consider Dean Smith, the Hall of Fame basketball coach from North Carolina. He spoke out about the Iraq war before it even began. Good Christian, Dean. And if you think that's an easy stand to take in a state like North Carolina, think again.

Then there's Mark Cuban—who's sponsoring the Fallen Patriot Fund. It's a nonpolitical group. But Cuban (love him or hate him) knows that American soldiers and their families are paying a terrible price, and he's decided to step forward and make a difference. Fallenpatriotfund.org.

And if you're thinking, Why should I listen to some stupid-ass elitist Hollywood lefty talk about what's best for my country, then don't. And go to Military Families Speak Out (mfso.org) and see what the brave people whose kids are on the front lines have to say.

The Web site headline that really grabbed my attention—"Bush Says Bring 'Em On, But We Say Bring Them Home, Now."

It's easy to talk tough, if it's not your kid who might be lost.

And it's my wife who reminds me, "Power for Shiites means burkhas for the women."

There arc parallels between sports and war. Ultimately, both are won by the folks who want it more. If the Iraqi people won't face down the thugs in their midst, then we sure can't do it for them.

It just won't work.

Where are their Jimmy Stewarts and Harriet Tubmans? Let their Gary Coopers step up. And Audie Murphys and Joan of Arcs. And Buffalo Soldiers.

I'm watching Iraqis cut and run as our guys get cut down. It ain't right.

The last time so many American lives were being lost, it took too many years for us to come to our senses.

And every life lost was a damn crying shame.

Here we go again.

Every life lost is a damn crying shame.

God bless America.

God bless humanity.

God bless common sense.

Let's go, Mets.

22

BASEBALL IS A CRUEL CRUEL GAME

I saw in *Variety* that Jerry Bruckheimer's finally gonna make *Glory Road*, a movie about the all-black starting five from Texas Western who defeated Adolph Rupp's all-white Kentucky Wildcats to win the 1966 NCAA championship.

Now that's a story worth telling.

Some have called Adolph Rupp a racist, others disagree. To me, one thing is very clear: As a college basketball coach in the '40s, '50s, and most of the '60s who refused to recruit a single African-American player, Rupp was either a racist or one of the stupidest guys on the planet, take your pick.

Glory Road. I gotta look into this. You know there's gonna be some serious basketball action in this movie.

And I need a job. (Every time I conveniently forget that fact, Providian and Cross Country and American Express are happy to remind me.)

The last basketball adviser job I had paid me $3,500 a week for four straight months. That's a lot of money for telling millionaires to quit hanging on the rims.

Glory Road. I'm not sure who's gonna be cast as Adolph Rupp (with any luck, Dick Cheney will soon be available). But Don Haskins, the Texas Western coach, is gonna be played by Ben Affleck.

Yeah, that Ben Affleck.

Now there's no reason to start with the cheap shots when Ben's been doing a pretty good job of it himself. He's just now staggering out of that Celebrity House of Mirrors, dazed, bleeding, eyes wide open in disbelief that the paparazzi flashbulbs have stopped popping.

And now he's gonna play Don Haskins. This absolutely shocks me. Not because Haskins was a seasoned, crusty good old boy and Affleck is a young Bostonian whose mother still cuts his French toast for him. And not because Haskins looked like a vulture on a fence post and Affleck looks like J-Lo's latest discarded boy toy.

No, I'm shocked because in 1966, Don Haskins was wearing flaming yellow and red checked sports jackets that even Andre 3000 wouldn't be seen in, while Ben leans toward Armani.

(I'm assuming it's Armani. What do I know? My taste in clothes parallels my taste in cars. As in, if I could buy a used Camry from JC Penney, I would.)

So, *Glory Road*. Could be a job for me.

Now I'm sure some of you who are reading the column are thinking, How does this smart-ass think Bruckheimer's ever gonna hire him when he just trashed his movie star?

Good question.

There's a guy named Harry Knowles who's got a movie Web site called ain't-it-cool-news.com. This guy had a penchant for sneaking into sneak previews and posting such scathing reviews that some bigwigs in Hollywood actually broke down and gave him a nice fat development deal in a thinly veiled attempt to buy his silence.

Maybe it'll work like that.

Besides, I didn't trash Ben Affleck. He trashed himself.

But maybe it's something bigger. Maybe it's a flaw in my character.

It's like how I have to keep pushing the envelope with my ESPN editors in Bristol. (Ah, they're groaning again.) They killed my column last week. Seems there was a slight misunderstanding. Like the late, great character actor Strother Martin said in *Cool Hand Luke*, "What we have here is a failure to communicate."

As in: Here's how much rein we're gonna give you. And not one inch more.

So when it comes to the column, I'm back on track. Cinch yanked tight, halter in place, blinders on, and spittin' the bit as I barrel down this week's backstretch. (Having at least made the case that geldings might run but they sure as hell can't write.)

The column readers had to write off last week's effort in which I solved the war in Iraq, fomented revolution against super-rich tax dodgers, called Mel Gibson the worst comedic actor on the planet, cried out for our troops, and made a case for polytheism—all in under 1,100 words.

Because what does all that have to do with sports and Hollywood?

"Absolutely nothing!" Or at least not enough.

So it's sports and Hollywood. Hollywood and sports. My new mantra.

Nam myoho renge kyo. Nam myoho sports and Hollywood.

Hey, how about those Iraqi soccer players that . . . uh, nah . . . forget it.

But look, to me life isn't about results, it's about process. The doing, the creating, the arguing. That's why moviemaking is such a fascinating endeavor. And that's why sports are so important.

Everybody getting all worked up about stuff and no one getting killed in the process. It's cool. You take a few lumps, have a few laughs. (Even at your own expense. Right, guys? Guys . . . ?)

Maybe Ben Affleck will make a good Don Haskins. Let's hope so, for the movie's sake.

But there are perils in casting. Or miscasting. Just ask the executives at Paramount who released that boxing movie, *Against the Ropes*.

I talked to someone close to the production. They'd put together a talented director, good script, and great surrounding cast (Omar Epps in particular), but for the longest time, they couldn't get it made.

Not until Meg Ryan committed to star.

But the audience didn't believe her as a gritty broad capable of handling the rough-and-tumble of the fight game, and the movie died a quick death.

How ironic. The star actually responsible for getting the film greenlit is the principal reason the movie tanks at the box office. It happens more often than anyone wants to admit. You gotta think, Why don't these people wise up? Why do they keep spending these huge salaries on "stars" who can't even open a movie?

Because there's a certain comfort factor that comes with the decision.

Meg Ryan has starred in movies that made gazillions of dollars. Ergo, if Meg Ryan stars in our movie . . .

The same thing can be said for writers. It drove me crazy when Kurt Rambis and I were doing the basketball coordinating on *Eddie*. We were in Charlotte, North Carolina, scrambling to pull things together when it became clear that the script was in dire need of a rewrite (if not all-out triage).

Hey, what about me? I knew the script well. I had clear ideas on how to fix it. I'd had writing deals at Fox and Warners with A-list directors like Ron Shelton and John Hughes, so it wasn't like I didn't have credentials.

I made my case, but no way. I was the basketball guy. I was pegged. It's the danger of wearing more than one hat. Besides, I was too cheap. That's right. My price for a rewrite hadn't risen into the stratosphere—proof that I wasn't up to the task.

What did the bigwigs do instead? What they always do. They hired these hotshot Hollywood rewrite guys to come in and fix it.

To the tune of $100,000 a week.

That's right, a week.

But when the first team only made things worse, they brought in some new rewrite guys. And the weekly fees climbed into the $125,000 range. A buck and a quarter in Hollywoodspeak. And the script got progressively worse.

There are about 30 of these highly paid screenwriters in Hollywood. It's an elite club. Since the making of *Eddie,* their weekly rates have climbed even higher as the studios keep going back to them, movie after movie.

Even when their efforts fail.

The comfort factor. It's the safest way for an executive to save his job.

"Look, I hired the best, most expensive rewrite guys in the business, so don't blame me if the movie sucks."

And who decides who are the best rewrite guys in the business? The Best Rewrite Guys in the Business Fairy.

So we all watched helplessly as *Eddie,* scene by scene, shot by shot, line by unfunny line, went straight into the toilet. And the rewrite guys flew off into the sunset, hundreds of thousands of dollars richer, on their way to the next assignment.

Leaving the rest of us to clean up the mess.

23

BUT YOU TWO MUST FAIL

The best athletes have a mean streak, a killer instinct.

Jim Brown. Roger Clemens. Marion Jones. Oscar De La Hoya. Luc Robitaille.

Isiah Thomas. I talked to Isiah a while back about the 1989 NBA Finals when the Pistons swept the Lakers, 4-0.

"How'd you guys pull that off?"

"We knocked them down, put a foot on their neck, and stuck a knife in their heart. Four straight games," he said, and smiled that choirboy smile of his. "They never knew what hit them."

This is the same smile that New York Knicks fans will soon grow to detest.

How do I know? I've got friends in the Continental Basketball Association, which Isiah single-handedly destroyed a few years ago. Greed and hubris. Killer instinct run amok.

But that attitude will serve you well in Hollywood. Out here, the competition is fierce. Every year film schools turn out thousands and thousands of ambitious young filmmakers.

Now let's be real—these people don't head for Des Moines upon graduation.

(Although they'll probably end up there. Producing commercials for the local merchants: "Ready, Mr. Whipple? And action!")

No, these young hotshot auteurs get out of film school and head straight for Hollywood. Throwing their scripts against the wall just like the rest of us.

And some of them are sticking.

Patty Jenkins, the young woman who wrote and directed *Monster*, with Charlize Theron, came right out of AFI and straight to the Academy Awards. My wildman screenwriter friend George taught her there.

"She had one movie that she was determined to get made," says George. "She grabbed this thing and wouldn't let go. She broke through to Charlize Theron, found the financing, and pulled it off."

It's an interesting strategy. Probably the smartest one to follow.

Single-minded mania.

There's a danger to having too many watches in that overcoat you're throwing open.

Go into a meeting, my best advice is to say, "Here's a movie that I am absolutely determined to get made."

If you start by giving executives choices, you're hurting your own cause.

Larry Gordon is one of the most seasoned movie producers in the business. I recently saw an interview where he said, "At the end of the day it's not who you are or who you know. It's the script. It's the story."

Talk like that makes screenwriters take heart. But still, you've got to fight through the thousands of others who know that they, not you, have written the next hot screenplay.

It's like sports. When I tried out for the freshman basketball team at Princeton (no scholarships, remember) there must have been 50 guys there. Our young coach, Artie Hyland, asked how many of us had been captains of our high school teams. At least 25 of us raised our hands.

We all looked at each other, some thinking—I was the captain of my high school varsity and there is a distinct statistical possibility I won't even get to become the twelfth man on an Ivy League freshman team.

While the rest of us, the actual dozen who made the team, were looking over the group thinking, Fuck these wimps.

It's like that. "I'm gonna make it, and you're not."

And know this—if you decide to give Hollywood a shot, you've got to deal with guys like Peter Bellwood.

Peter's a silver-haired Brit who came up through improvisational theater (*Beyond the Fringe, The Establishment*). He had a great run in London and New York, then wrote *Highlander* (a very trippy time-traveling sword-and-sorcerer flick) and found his way west to plunge into the life of a Hollywood screenwriter.

For a while there, Peter had the questionable fortune of being married to the television star Pamela Bellwood. For those of you under 30, Pamela Bellwood was the Pamela Anderson of the '80s, only way more (fill in the blank).

Peter and I were working on a script together. One day, we ran into a wife-husband writing team in a coffee shop. They'd recently had a script produced by HBO and we offered the customary congratulations. They were gracious. They asked how I was doing and I said fine. They asked Peter how he was doing, and he answered (in his grandest English accent).

"Fine as well. But you must understand that for me to be truly happy, not only must I succeed, but you two must fail."

We all laughed. But the lucky couple looked a little skittish as we walked away.

Get it? Sports and Hollywood. Sports—for every winner, there's a loser. Hollywood—for every winner, a thousand losers.

A thousand scripts in the crapper.

A thousand actors on the bus back to Atlanta.

Killer instinct.

I'm still looking for a director for *94 Feet*. A director with credibility who hasn't hit the big time yet (meaning he's still approachable). Plus a guy who really understands hoops and has some real maverick in him.

Paul Johansson fits the bill. He's got the confidence, the swagger of a smart jock.

We're sitting at Jerry's Deli on Beverly Boulevard, wolfing down pastrami sandwiches, talking through the challenge of making a movie about a single game. A movie that shows each team's point of view (POV in script language).

I relate what Ron Shelton says again and again: "Show the audience the game they can't see from the stands. Show it from the inside."

When you're in a game and you're exhausted and the score is close and the crowd is screaming, it can get very weird out there.

Time can speed up or slow down. The rim can look as big as the ocean or as small as a sand dab.

How do you show what it's like when you're guarding their best shooter and he's quick and moving without the ball and you're fighting your way through staggered screens?

Johansson talks about putting a handheld camera right on the floor. You make the cameraman the defender. You put the audience right inside the action.

We riff on other techniques. The use of light, of sound. How sometimes during a game a screaming crowd can grow abstract. How you can bleed out the image, warp the volume.

Moments like this, sitting with a director, sharing ideas, getting excited, this is Hollywood at its best.

I don't know if Johansson will end up being the director. But I am absolutely determined to get this movie made.

And determined to enjoy the process along the way.

There's this Bulgarian artist Christo who's famous for creating huge environmental pieces like *Running Fence* and *The Curtain at Rifle Gap.* He wraps buildings and landscapes with miles of colorful fabric. It's great stuff.

But to get it done, he and his team have to deal with ranchers and environmentalists, school boards and zoning regulators. It's one hassle after another.

Someone asked him, "Don't you wish you were simply free to make your art? To not have to go through all this difficulty?"

Christo's answer, "No, no. The difficulty is the art. The process is the art."

The process.

The process of work. The process of life.

The process of grinding out a couple thousand words a week while trying to salvage a so-called career. A sudden shout from my wife. Our six-year-old has somehow lost a stick of butter. That's a first. That's my process for the next 30 minutes as we turn the house upside down.

Sadly, this process of work and life is one that more and more young Americans will never have the chance to experience.

Pat Tillman was killed in Afghanistan. The media are making a real circus of it. It smells fishy. It smells like heavy-handed propaganda. People missing the point. So in response, I sneak this one by my editors at ESPN.com.

Last week a former football player was killed overseas while serving his country. He was caught in a firefight while coming to the aid of his sergeant.

He was recently married and his young widow just learned she is pregnant. This brave football player hailed from West Virginia and his obituary appeared in the *Wheeling News Register.*

And no where else.

He was Lance Cpl. Michael J. Smith Jr.

R.I.P.

24

TOMMY LEE JONESTOWN

Chances are if you're not a father already, you're gonna be, so listen up.

(Apologies to my six women readers. You too, Mom.)

Guys, never coach your own kid.

Not in games, not in practice, not in the backyard.

If your kid's out there, gripping the baseball bat bass-ackwards, don't try to correct him. He'll argue that this is the way the real pros do it and what do you know anyway.

Just pitch him the damn ball and see what happens.

Or better yet, sign him up for soccer.

Because baseball is a cruel cruel game.

Ask Derek Jeter, who once went 0 for 32. Ask Bill Buckner—unfairly blamed for losing the 1986 World Series. Ask me—still haunted by a dropped fly ball in Little League. Still resentful after working with Tommy Lee Jones on *Cobb*.

In baseball, you're all alone out there. Unlike other team sports, you can't get lost in the crowd. (That's why soccer is so popular among P.C. parents. "Isn't it beautiful? They're all out there being mediocre together!")

But baseball, jeeez.

My 12-year-old, Cole, approached me about playing hardball this spring for the first time ever.

"Aw, man, you don't know what you're asking for," I said.

"I want to play."

Cole's developed that winner quality as an athlete, mostly because I've stayed out of his way. He can take a hit. He's shocked when he drops a pass. He wants the basketball in his hands as the clock winds down.

What's baseball gonna do to that?

"Only if I can find you the right coach," I told him.

In youth sports, that's "THE BIG IF." (Is there a movie here?)

Because these guys can be real knuckle-scrapers.

Bullies, alcoholics, pill-poppers, frustrated ex-jocks with axes to grind. Men under the thumbs of their bosses and/or ex-wives.

So you sign your kid up thinking, Okay, he can handle a Walter Matthau type. It'll toughen him up. But what if he ends up with Tommy Lee Jones on a bad day? What then?

I've got a buddy whose kid, James, pitches in Pony League. They were late to a practice, late because James had cello practice.

Next thing the coach is yelling, "Cello!? You play cello!? What kind of wuss are you!?"

This is a grown man talking to a 12-year-old boy. My friend isn't happy about it, but his son doesn't want him interfering.

Cole was playing club basketball for a coach who was a real screamer. I went up to the guy.

"Why are you yelling so much?"

"How else are they gonna know it matters?" he answered. "I'm yelling to show the kids how much I care."

And this guy wonders why he can't keep a girlfriend. ("Jimmy, was that foreplay or a suicide drill?")

I tell the guy, "I don't mind you barking at my kid, but don't scream at him."

One of the great sprinters (Michael Johnson, I think) was once asked why didn't he ever go out for football—you know, make like Herschel Walker.

His answer—"I watched a practice. Too much yelling goin' on. All these coaches yelling and screaming. That's not my scene."

So there's that, and then, of course, there's the issue no one wants to discuss.

My sister-in-law, Sandy, is an executive producer for NBC's *Dateline*. She did a piece about a Little League coach who molested dozens and dozens of boys over two decades. He would take months, even years, to ingratiate himself with the parents. They had no clue. He ate at their dinner tables and molested their sons.

Sandy says, "The scariest thing? This was the nicest guy on the earth."

What kind of crappy-ass world is this?

Sorry about this bummer tangent, but let me finish.

Look, for all you wonderful youth coaches out there—the ones doing it right—keep on keeping on. And please don't take it personally when a parent doesn't want their kid to be alone in the car with you.

For all you parents of young athletes. Be vigilant.

Okay, enough. We were talking baseball.

It was 1993. Ron Shelton asked me to help out on *Cobb*, a movie he was directing for Warner Bros. about baseball Hall-of-Famer Ty Cobb.

I didn't particularly want the job. I'm not an ultrasavvy baseball guy. I didn't want to have to fake things. But Ron assured me that he'd cover the technical stuff (having played Triple A baseball for years) and my job would be more logistical—hire players, arrange practices, like that.

I still had doubts. I'd heard that Tommy Lee Jones could be a real handful. And the guy he was playing, Ty Cobb, had been just plain cruel.

The notion of spending the next 12 weeks with these two personalities rolled into one sent me on a search through my 14 credit cards.

Sorry. I had to take the job.

We started prepping out of Ron's office on Sunset Boulevard in Hollywood.

Tommy Lee Jones was in town but I hadn't met him yet.

I'd been scrambling all day, breaking down the script to see how many players we'd need, searching for a double for Tommy, finding a local practice field to go over some stuff.

I'd gotten back to the production office and had sprawled out on the couch. Next thing you know, Mr. Jones is storming into the office yelling at me.

"Wake up, boy! Be somebody!"

"Okay," I said, sitting up.

But before I could introduce myself, he was in Shelton's office, doing the hearty hello thing. So that's how it was gonna be. Above the line, below the line. That was okay. There's no sense in trying to be these guys' friend.

We ended up in Poinsettia Park that afternoon. Ron, Tommy, Manny (an excellent double for Tommy except for his hat size), me and a couple other guys.

Ron was demonstrating how Ty Cobb held a bat with a split grip. He was a lefty, besides. The right-handed Tommy grabbed the bat and in short order had a credible left-handed swing going.

The guy was a solid athlete, no question.

Then we moved over to the third-base line. Sliding practice. We asked Manny to demonstrate several types of slides. Hook, headfirst, cleats up. The dirt was hard. Manny was getting pretty beat up, but he was game. Tommy watched intently.

"Okay, I'll work on that back at my ranch over the next couple weeks."

Later, Manny pulled me aside. "You oughta warn him how dangerous sliding is. Lots of injuries."

I thought about it—he was right. So I found a moment and said, "Tommy, back on your ranch, be careful. Sliding can be dangerous."

"Dangerous? Sliding? Don't be stupid!" he yelled. "I'll tell you what's dangerous. Falling off a horse, that's dangerous!"

I shrugged.

So Tommy Lee Jones headed back to Texas and I headed to Detroit, where I'd spend the next two weeks hiring dozens of players, meeting with stadium personnel and local production crew, prepping for scenes that would never be shot.

This is not unusual. In the movie business, there's always something.

Then I got the call at my hotel. Tommy Lee Jones had broken his ankle practicing sliding on his ranch in Texas. The whole production was turned topsy-turvy. They reworked the entire schedule. They wrote out all the scenes at Tiger Stadium, and we ended up in Birmingham, Alabama, to shoot the baseball.

This wasn't an easy movie—a difficult actor playing a difficult man. But Shelton had a firm hand on the reins. Director as coach. Ron doesn't tolerate any crap. He's got a great way of yelling without yelling:

"You know, I can yell! I can get really pissed off! Anybody want to see that? I didn't think so! So let's tighten it up around here!"

Tommy Lee Jones had just gotten his cast off. He and I quickly established a perfect working relationship. We didn't say a word to each other unless it was absolutely necessary.

I've found that passive-aggression is the most useful weapon in an ongoing battle with your superiors.

What're they gonna do, fire you for not saying hello?

So there we were, manly men shooting baseball.

Then Roger Clemens came down to give us a few days as a rival pitcher, and God, the testosterone really started flying.

Clemens was a pro's pro. Cordial but curt. All business. His warm-up was something to witness. He'd start by throwing from 20 feet behind the mound, then he'd slowly move up. He said by the time he got to the rubber, the strike zone looked as big as a house.

Players were scrambling to catch for him. The crew guys were in awe. Grips who wouldn't turn their heads at the sight of a topless starlet were standing there, mouths open.

I strolled over and stood alongside the catcher. It was unbelievable. The ball coming in like a missile. Hitting the catcher's glove with a sharp *POW*.

Clemens stopped for a second and grinned at me. "Hey, why don't you step in. Come on, grab a bat."

"No way," I shouted back.

"Come on," he yelled, laughing.

"How about this?" I laughed back. "No *fucking* way!"

Clemens smiled. Did he respect me less? I don't think so. He knew the power of his stuff.

You may have heard me trash basketball players who think they're so tough. And how football, hockey, and boxing are the sports where you'll find the real tough guys.

I never mentioned baseball. But checking out Clemens close up and considering how sometimes pitchers purposely hit batters—you've got to be damn brave to step into that box.

I'd rather face Warren Sapp.

Or Tommy Lee Jones.

He stayed in character the whole shoot.

One day we were out by second base under the hot sun. The great Aussie director of photography Russell Boyd (who later won an Academy Award for *Master and Commander*) was setting the cameras.

He was a quiet, unassuming man. He did a brilliant job shooting *White Men Can't Jump* for a reason you'd never consider.

"Oh, my gawd!" he said when he first saw Wesley Snipes and Woody Harrelson together. "How am I going to light you two?!"

Remember, black absorbs light while white reflects it. And you can't get much blacker or whiter than Wesley and Woody. A D.P.'s worst nightmare, but Russell pulled it off.

Anyway, out there by second base on *Cobb* we were facing a delay. Camera problem. Shelton headed back to his trailer to review the schedule.

Unfortunately, Tommy Lee Jones had already been escorted out to the set. He sat there in his director's chair, fuming, still in character. And that Ty Cobb needle came out as he went after Russell Boyd.

"Hey, Russell, what the fuck is this? I'm sittin' here in the hot sun watching you muck around like some schoolboy?"

Russell said nothing. Kept right on working. There must have been 25 of us or so, quietly standing, milling about. Holding our breaths.

Tommy continued. "It's a camera, Russell. You look through the viewfinder. You push a button. Hey, you got film in there, right? Russell, I'm talkin' to you."

God, it got uncomfortable. The unflappable Russell Boyd continued ignoring the insults.

Sometimes it hurts worse than getting picked on yourself. When a real good guy, the kind of man this world needs more of, is subjected to this crap.

I felt like saying, "Hey, Russell, how many awards did you win for all that brilliant work you did with Peter Weir? You know, *The Last Wave. Picnic at Hanging Rock. The Year of Living Dangerously.* All that other great work you've done?"

But did I? No. I stood there, fuming, until Mr. Jones had satisfied his urge to humiliate.

Ty Cobb was a great baseball player.

Tommy Lee Jones is an excellent actor.

Baseball is a cruel cruel game.

I snap back to Pony League. It's the bottom of the seventh. Cole is 0 for 3, but at least he made contact. He'll survive.

The lanky, quiet, cello-playing James is at the plate. He hasn't played well. His coach has been on him.

There's a kid on first. They're down a run. James smacks a towering fly ball straight over Cole's head in center field to win the game.

He crosses home plate triumphantly. His coach is right there to high-five him (undoubtedly convinced it was his needling which sparked the homer).

As they head for the bench, I shout out through the chain-link fence.

"Hey, James, that's what you get for playing the cello!"

The coach shoots me a look. James smiles.

In a world full of big-time bullies, it's the small victories that mean the most.

25

FROM PIPE DREAM TO PAYCHECK

They say don't explain or complain, but what do they know? I woke up this morning cryin' boo hoo hoo, I can't take this freelance life for one more day.

Stick me in a cubicle. Make room at the assembly line.

I once got a job at the Princeton University Store sticking price tags on erasers. I lasted about two hours. That's the job I want right now. Or how about cutting lawns? Cutting lawns on a golf course, that's the job for me.

But noooooo. I'm the guy who pledged that I would *never* work for "The Man." (I never mentioned "The Mouse.")

Maybe I'll start my own business—"ZenMow. Our blades are sharper than our minds."

You know you're on a slippery slope when you're trying to get your six-year-old to feel sorry for you.

Not rock bottom. But slippery. I've been closer. Writers' strike, 1988. I'd really been catching my rhythm—repped by William Morris, a handful of deals under my belt. Then we went on strike for five months, and it all came crashing to a halt.

Hanging with some screenwriting pals, drinking at Boardner's on Cherokee, right off Hollywood Boulevard. Suddenly I get that queasy feeling and somebody's asking, "Hey, man, why do you look so green?"

I stagger out to the sidewalk, kneel down behind a parked car, and lose my dinner, lunch, and breakfast (in that order). Then I roll over onto my side and just lie there in the dark.

Suddenly a voice.

"Hey, buddy, you all right?"

I peer into a boarded-up storefront. A homeless guy's sitting there, all ratted out, filthy-skanky, long matted hair.

"Yeah, I'm uh, I'm okay . . . ," I say.

"Anything I can do, just let me know," the guy answers.

Now that might not be rock bottom, but it's close.

Have I mentioned there's another writers' strike looming? These things can get very dangerous, because most WGA writers aren't working anyway. Vote to strike, you won't have to keep explaining why you don't have a job.

So back in the present, still feeling sorry for myself, I wonder what my father would say if he were alive right now.

"Tough—you want to know what's tough? Running down a blind alley in Karbalā, Iraq, with 65 pounds of gear in 100 degree heat with RPGs flying every which way and buddies getting blown up and you're thinking, Aren't the Shiites supposed to be on our side? That's tough."

My father worked for UNEF, the United Nations Emergency Force. The Belgian Congo, the Gaza Strip, Cyprus, Lebanon. He knew a thing or two.

It's hard to make someone proud when he's dead (something to keep in mind if your old man's still alive).

I suck it up and get to work.

As a freelancer, you have to prioritize. What's real? What's the closest paycheck? What do I have to get done? What can wait? What's a pipe dream? When do I cut my losses?

My pal Clay Moser was an assistant coach to Jimmy Valvano at N.C. State, then spent years in the CBA, coaching and running organizations. A top-notch guy. He explains the method of a CBA owner, a supermarket entrepreneur he worked for.

"This guy carried a three-by-five card in his shirt pocket. A list of the six or eight things he absolutely had to do that day. He got straight to work and crossed them off, one by one."

Not something they teach at Harvard Business School. Too simple.

So I'm staring at today's three-by-five card.

1. **Call *Glory Road* line producer.**
2. **Answer Ron Shelton email.**
3. **Answer 103 readers' email.**
4. **Contact Baron Davis re *94 Feet*.**
5. **Work on book proposal.**
6. **Complete reality show one-pagers.**
7. **Write column.**

Okay, here we go.

1. CALL GLORY *ROAD* LINE PRODUCER.

It's a basketball movie which just starting prepping—Ben Affleck playing a college coach. A Jerry Bruckheimer production.

I make the call and get through to Dana, a production coordinator. I'm looking for Andy Given, the line producer. This is my fourth call. They've had my resume for three weeks.

Dana says, "Not to put you off, but there's a whole stack of resumes and he hasn't gotten to them yet."

"I understand."

"Plus it's a location shoot, and he'll probably be doing most of the hiring locally."

(For other jobs, yeah, but how many experienced basketball movie guys is he gonna find in El Paso?)

"Did you get a start date?" I ask.

"August."

"Great, thanks, Dana. And please remind him, whether he hires me or not, it's critical to get an early start on the uniforms. They're always a nightmare."

"Okay, I'll tell him."

I hang up thinking, August, that's good. Gives me some time here.

This would be a cool job.

Ben Affleck's playing Don Haskins, the coach of Texas Western who fielded an all-black starting five. They beat all-white Kentucky to win the 1966 NCAA Championship.

That's dramatic stuff.

I get an email. Saul Smith, who recently played for his father, Tubby, at the University of Kentucky, is interested in the movie. Now there's an ace up my sleeve. If I get the job, I'd like to bring Saul Smith along.

He'd really help grease the wheels with the Kentucky administration. Help with the players. Game film, uniforms even. And provide some great pub, since he and his father are African-American, and there they've been, center court, at a once segregationist university.

(You know, through all the crap and bad-mouthing, this country has made some great strides, and we all ought to be proud.)

The toughest problem on *Glory Road*—how do you find a bunch of hip-hop nation ballers willing to wear those itty-bitty basketball shorts? You can just imagine the first time these guys show up in wardrobe.

"Ooooh, man, how much you payin' us?"

Okay, the call's been made. One is done.

2. ANSWER RON SHELTON EMAIL.

Shelton's had some great successes and taken some hard knocks. He spent years playing Triple A baseball in the Baltimore organization back in the days before free agency. There were a dozen teams in the major leagues he could've played for, but the Baltimore lineup was a tough nut to crack. So he finally walked away.

Then turned around and wrote and directed *Bull Durham*.

Like I said, ups and downs.

Last week in my column, I spilled some serious beans about working on *Cobb* with Tommy Lee Jones and what a mean guy he could be.

This was dicey stuff, but Ron was cool with it. He emailed me, "In many ways Tommy Lee was one of the easiest actors I ever worked with, and I'd love to do it again. I know he ran roughshod over everyone around him and ate p.a.s for snacks, but in terms of preparation I've rarely worked with an actor who was more ready, and in that sense, respectful of the director."

A few weeks ago I was flipping through the channels and came across *Coal Miner's Daughter*. Sissy Spacek and Tommy Lee Jones. 1980. It was the scene where Loretta Lynn's husband (played by Tommy) is showing her the property he bought to build their dream house.

He's all full of himself, got the house all laid out. Loretta is unexpectedly pissed, and the two characters thrust and parry their way through a whole range of emotions—indignation, surprise, frustration, explanation, reconciliation.

It's a beautiful scene. Wonderfully written. Cleanly directed.

And perfectly acted.

I mention the scene to my wildman screenwriting pal George who teaches at AFI. George's eyes light up.

"Gil Dennis teaches that exact scene. Every fall, to the entire class of AFI fellows entering that year."

So, if you're interested in the art and craft of moviemaking, rent *Coal Miner's Daughter*. Fast-forward to the mountain meadow. And savor some of the finest film acting you'll ever experience.

It's enough to make you go, "So what if he was a pain, it was an honor to work with a talent like that."

But still, in my book, human kindness rules. And there are plenty of guys who manage to stay gracious all the way through. Steve Martin was a prince. Same for Nick Nolte. Lou Gossett Jr., Dan Aykroyd. Diane Keaton, and Mary McDonnell.

Back to Shelton's email. I respond, taking the opportunity to ask a favor. J.C., a producer with rights to a very funny college football novel, wants me to write the adaptation. I ask Ron if he'll be a "friend of the project." Offer some guidance through the writing process. It's not a request I make lightly. Even with friends (especially with friends), there are only so many chits you've got.

He agrees to do it.

That's gold to me. A writer-director of that caliber, available to call upon? That's true value. Wealth you can't wear around your neck. That's enough to make me want to call it a day.

But the three-by-five card says no. Five more to go.

But hey, I'm through with two.

3. ANSWER 103 READERS' EMAILS.

I power through them, appreciative of all the feedback. I'm somehow gaining quite an eclectic group. More women than I expected. New friends from Italy, Sweden, and the Philippines. A new friend in a wheelchair. Guys named Leroy and Carlos. Guys stuck in cubicles. People looking for a different way to live their lives. Or looking for a cheap laugh.

Hollywood people, executives, film students. Youth coaches. Athletes. Couch potatoes. Writers (we're all writers). Sports lovers.

There are a lot of conservative readers calling me out (lefty that I am). It's a cool, respectful dynamic. We often agree to disagree.

The further left you travel, the sooner you're heading back full circle. Look out, all you libertarians, here I come. H. Ross Perot and Ralph Nader have more in common than you'd ever realize.

The day I stop talking to my conservative friends is the day you can take me out back and shoot me. How about we all hang out together on the far side?

I finally knock out the last email.

Okay, free of three.

4. CONTACT BARON DAVIS RE 94 FEET.

Baron Davis is financing several movie projects, and he's interested in my college hoops movie. The Hornets just got knocked out of the playoffs. I leave a message with his agent, Todd Ramasar, but say no hurry, give him a couple of days to get settled back in L.A.

So four hits the floor.

5. WORK ON BOOK PROPOSAL.

Aw, man. This one's killing me. I know there's a book in this weekly journal I'm keeping, but my agent, Matthew Guma, says it's not enough to just slap them together. He wants me to clean them up, turn them into a real book. I tell him, "Screw it. I don't have the time. I'd probably just make it worse anyway."

Just the idea of taking certain stories and organizing them by theme and blending them all seamlessly together into some easily digested smoothie—*Hollywood Jock Lite*—sorry, I just can't do it.

Besides, how unfair would that be to my regular weekly readers? All those folks slogging through my tortured syntax, lurching transitions, and endless weird tangents.

You are the greatest. The lonely and the brave. (The bored and the witless?)

Why should I make it easier on everyone else?

I tell Guma, "Think of this book as a long, slow-motion train wreck. I'll add a little commentary in between the columns and call it a day. Oh, I'll also plug in some of the pithiest emails I've received."

"Like what?" he asks.

"Like this," I answer:

"Wow, racism AND sexism all in your first column. Can't wait for all the other 'isms' that may follow. Oh, yeah, suckism was in there too."

See? A total of 138,000 hits in the first 12 hours. I ask Guma, "Why mess with success?" He answers, "Doh . . . to get an advance, perhaps?"

I tell him I'll think it over, hang up, and write a short intro for the book of columns.

Still alive after five.

6. COMPLETE REALITY SHOW ONE-PAGERS.

I get a lot of email from people with reality show ideas. First, if you're serious, don't share your ideas without registering them: www.wga.org. Second, know that five or six major producers have pretty much locked up the market, and the chance of your getting a show produced is virtually zilch. Third, if it doesn't involve hot tubs and/or humiliation you're really sunk.

Still, I've got two ideas that I gotta put out there.

The first is the stupidest and cheapest, so probably has the best chance of getting made.

Color-Blind Date.

Set in steamy Los Angeles, the multicultural capital of the universe.

I wanna see Samoan gangstas dating Valley girls. Koreans mixing it up with Nigerians. Throw in meet the parents, have them all end up in a hot tub, and ladies and gentlemen, we have a winner.

I'll recruit my friends Pookey (African-American), Rasha (Egyptian-American), and Maya (Indonesian-American) to join me (Polish-Norwegian-American) to help produce. (I love L.A.)

My second idea I'm taking more seriously. *Mind Your Own Business.*

This is in response to Donald Trump's *The Apprentice.*

A competition among real entrepreneurs (drawn from the more than 35 million, all across America) who are starting—or looking to grow—their own businesses.

I've spent the last several years developing a summer four-on-four full-court pro basketball league (we're eventually gonna get this up and running). Along the way, I've learned a thing or two about how tough it is to start a business.

This show will be the real deal. None of this pouffy-haired, richy-rich brownnosing B.S. I'm talking businesses like hair salons and bowling alleys; delicatessens, comic book stores, bakeries, and pet-grooming services.

Watch these intrepid American entrepreneurs compete for start-up capital and second-stage financing. This is good stuff. Hands-on, educational, lively.

The ultimate competition, because it's real life.

You don't get to go home if you lose. You might not have a home.

Get Warren Buffet to host. Or how about H. Ross Perot? Don King? Or H. Ross Perot *and* Don King—now we're talking.

Put it on, and watch it catch fire.

Okay, two hours later, I've got a one-pager knocked out on each.

Number six is in the mix.

7. WRITE COLUMN.

Consider it done.

I'm signing off. Another day in the life of a Hollywood jock.

Say a prayer for our troops.

26

GEORGE AND ROB'S EXCELLENT ADVENTURE

Do you know that journalists and columnists don't write their own headlines? It's true. So next time you see a really stupid headline, don't blame the poor schmuck who put the actual work in. Blame his boss. (I'm pissed about how my new Page 3 editor at ESPN.com is treating me. Can you tell?)

It's almost midnight, and I finally feel that necessary jolt of panic that drives me to the keyboard. God, why does it have to be like this?

George and I are pitching *Zulu Wave* at John Goldwyn's company at Paramount tomorrow morning and there's no way we're ready.

A young black surfer in South Africa breaks the chains of apartheid.

It's great stuff, true stories, vivid images. Black teenagers surfing under a full moon. Thousands of Zulus walking onto the all-white beaches in silent protest. Batons, whips, police dogs, massive waves, sharks—rage, courage, action, victory.

George's writing partner, Mark, came back from Durban brimming over with these tales. They enlisted my efforts and for the last several weeks we've been building the story beat by beat.

It's prickly, dynamic work. Email outlines flying back and forth. Mark and George have been doing the heavy lifting—while I've been jumping in with suggestions, new characters, possible reveals.

It's come together—we've got a strong, eight-page outline.

We're still not ready to pitch, but this meeting presented itself, and we figured what the hell.

One of George's students at AFI, Jamal, interns for John Goldwyn at Paramount. Goldwyn until recently was president of Paramount Pictures, but a long string of money-losing movies finally brought him down.

In Hollywood, when top executives are shown the door, they're given producing deals, nice offices, and a chance to make their own movies, right there on the lot.

But it can be a real jolt. Suddenly these same people who've been playing God (or at least Nero—you know, doing the thumbs-up, thumbs-down routine) find themselves under someone else's thumb, just like the rest of us.

We're pitching to a guy named Franklin Leonard, Goldwyn's head of development.

When you pitch like this, you're expected to show up with a "leave-behind," a page or two, some sort of written explanation of your story. These things are a bitch to write.

If they're too detailed (like our outline) you're just handing them stuff to nitpick apart, reasons to pass. But it's gotta be more than just a log line, more than just a one-sentence description.

A one-pager, that's what's needed.

And given George and Mark's hard work, this one's on me. So finally, nearing midnight, I ask myself, What is this movie about? The answer comes and I pound it out (three quarters of a page, 160 words). I email it off to George.

He's still up. He likes it. We talk logistics about tomorrow morning's meeting at Paramount. Did he confirm? Yes. Will there be a drive-on at the gate? Yes. We talk transportation.

George drives a battered Mercedes station wagon that makes my tired old Camry look almost perky. Ah, the personal sacrifice a grown man will make for his kids. For trumpet lessons and surfboards, baseball gloves and books and concert tickets and air-soft BB guns, for art supplies and club soccer and new shoes and more books and more music and McDonald's and Subway and gallons and gallons of milk, and barbells and headphones and you name it, they need it, and one way or another, they're gonna wheedle it out of you. Because they deserve it.

For our boys, George and I drive crappy cars. Proudly.

But who wants to show up at a studio gate in a beater belching black smoke? So once again, I borrow my wife's shiny white minivan.

It's the next morning, and I'm running late. George and his wife, Liz, are standing at their gate. George jumps in and Liz simply waves. It's a brave, wistful wave, the kind my wife saw me off with just minutes ago.

If waves could talk, this one would say:

"You'd better come through here, big boy, because there was this guy back in college, and he's making a nice steady six-figure salary right now, living in a big white house on a tree-lined street, yet I chose you over him. And that man, he loved me."

"Not the way I love you, baby. Not like me."

"That's for sure."

Cranking along the freeway, George is boiling over with energy. A jumble of ideas, stories, a nonstop rant that quickly has my neural transmitters short-circuiting.

First there's his weekend writing project, *Screen Dogs*—screenwriter as five-dollar hooker, a toothless wonder, working under the freeway overpass. Next thing I know, he's ranting on about a guy so plagued by his past that he resorts to sitting on the bottom of his swimming pool in full scuba gear.

I throw a monkey wrench at him, but he ducks and launches into a rant about his ancestors—manic-depressive Polish aristocrats, knocking around gloomy mansions, staring out into the rain, fingers drumming the windowsills, living lives of loaded revolvers and mental illness before lurching across the ocean into a hardscrabble existence in Jersey City. . . .

And I'm thinking, What am I doing here? I've got half a dozen other projects—promising stuff, with bona fide movie and TV producers, big-time book editors, and here I am, grinding along in my wife's Honda Odyssey with a Polish-American lunatic.

But deep down, I know George is the real deal. And *Zulu Wave* has got its hooks into me.

I try again to steer him back to the task at hand but he's onto his boys, Dash and Tristan, and their shizzle to the gizzle and their brawls over the bathroom, and Dash his oldest, roaming the yard blasting Miles Davis on his trumpet and Tristan, the sixth-grader with the size 12 feet, and how he cut the cheese at the dinner table, and Liz got pissed but George said, "Hey, T, good one."

And Tristan answered, "Willy Weedwhacker," and broke up.

And suddenly I'm laughing, weaving the minivan through the thickening freeway traffic, helpless against this latest onslaught.

"Willy Weedwhacker!?"

"They name 'em!" yells George. "Fartizzle to the sizzle. Sneaky Pete. Flatfoot Freddy."

By now we're both howling, me trying not to crash the car. "All right, you got me, I give up!"

Women hate this stuff, but for guys . . . Hey, I guess you gotta be one to understand.

I finally catch my breath and say, "George, how about we talk some *Zulu Wave* here?"

"Walter the Whiner. Henry the Fourth."

"You're drivin' me crazy!"

"Okay, man, sorry. Sorry."

"In half an hour, we gotta pitch this thing, remember?"

"All right, here's the pitch," says George.

I glance at the dashboard clock. It's 10:37. If this pitch lasts more than 10 minutes, I'm pulling the plug. These things gotta be quick. "All right, go," I say.

And that fast, George launches into a powerful, emotional speech. It's the leave-behind I knocked out last night, word-perfect (I mean every single word memorized) but delivered extemporaneously, with amazing cadence and conviction.

Remember in *Pulp Fiction* Samuel Jackson giving that biblical speech from Ezekiel to those Flock of Seagull, bad-guy-wannabes? It's like that. George is that good.

And then he's done. I glance at the clock—10:41. Four minutes and change.

"Jesus," I say, thinking, This guy's a goddamn genius. "That was impressive."

"Yeah? Your words."

"Our words," I say. "Mark, you, and me. I just condensed them."

We talk it over, decide to add a detail or two, a beat at the end with the surfer's father.

"All right," says George. "I got it."

George's cell rings. It's Jamal. "Where the fuck are you guys?"

"What!? You said 11 o'clock."

"Ten-thirty," says Jamal.

I groan. How many times do these meetings get messed up? You gotta double-check, triple-check.

George yells into his cell, "We're just turning onto Melrose. We're like five minutes away."

He snaps his cell phone shut. "That's messed up. Look, Jamal says this guy Franklin's got an eleven o'clock with Goldwyn and some other writers. We might have to reschedule."

"Fine, no problem. But let's try to get in for ten minutes."

I remind George that we're just testing it here. The odds of John Goldwyn buying this thing are pretty slim. There are other places to go. The producers of *Bend It Like Beckham* are expressing interest. The veteran English director, John Irvin, wants to hear more. There's Miramax and Sony Classics. HBO and Showtime. Fox Television Pictures. MGM. New Line. Focus Films. A couple dozen possibilities. How about Charlize Theron? She's South African and Oscar-hot, we'll get it to her.

I swing the minivan off Melrose into a long line of cars crawling through the Paramount gate. Security is out in force. Guys with clipboards, headsets, long-handled mirrors checking under the vehicles.

They ask who we are, where we're going. One guy takes our driver's licenses and disappears into a guard shack.

It's taking forever and the clock is ticking. I don't want to come back, I want to get in and pitch this thing. We're bouncing off the ceiling, "Come on, come on!"

Another guard eyeballs us and nods to his pal.

It's not al Qaeda the studios have to worry about, it's guys like George and me—crazed, grizzled screenwriters with bones to pick, wrongs to right. They see it in our eyes.

Facial profiling.

The guy with the mirror spends an inordinately long time working the underbelly of the Honda Odyssey. "Come on!"

We finally get waved through, park, and hustle over to the Maurice Chevalier Building. (For those of you under a hundred who don't remember Maurice Chevalier, don't worry about it.)

Jamal meets us at the door.

Franklin steps out of his office. He's young, African-American, dreadlocks reaching his shoulders. We all shake hands. He explains the situation. We tell him we only need five minutes. He ushers us in.

We all sit except for George. I say, "George is gonna do his thing here."

"Cool," says Franklin.

George goes to work, pacing the room, making eye contact, speaking forcefully:

"South Africa, 1987. A state of emergency. Riots in the townships. Protests in the capital.

"But for 15-year-old Kwezi Sisulu, it's the ocean that calls.

"Against his father's wishes, Kwezi's been surfing on a beat-up board for years. This kid is pure poetry. And now it's time for him to test himself against the privileged, hotshot, white surfers. But the beaches are segregated, the best breaks belonging to the whites.

"Young Kwezi decides to challenge the system. For this he pays a price.

"He's beaten, his board is broken, his father fired, his brother murdered, but still, Kwezi will not be denied. And finally, with the support of his Zulu tribespeople and a handful of progressive whites, Kwezi makes one final push.

"Picture the last scene at the Junior Championships, Kwezi and his father, touching foreheads in the Zulu way, as the father says, '*Amandla awethu!* The power is ours!'

"Kwezi enters the water, digs hard through the crashing surf. Reaches the lineup. Then a massive swell raises his board—and here a great reveal: thousands of black people quietly filling the all-white beach, there to bear witness as Kwezi tears it up.

"*Zulu Wave* is an uplifting, action-packed story of personal bravery, collective righteousness, and the ability of a lone, skinny-shouldered teenage boy to affect the tide of history."

George is finished. Franklin is quiet. Then says he likes it. He'll talk to Goldwyn. We hand him our leave-behind, shake hands, and we're out.

Amandla awethu.

27

THE SLOW DISINTEGRATION OF A GOOD IDEA

Unbelievably, some of the visitors to ESPN.com have actually been tracking my progress—offering advice, encouragement, contacts (and condolences). I've heard from guys at MTV and VH1. A guy starting a new PR firm. Producer Jerry Weintraub's nephew in New Jersey. Shaun Tomson, former world champion surfer from South Africa. Executives at Fox and Disney. Old girlfriends from high school. It's too cool. But I'm also living with the daily dread of having the plug pulled. How long are they gonna let me get away with hawking my own projects on their Web site? I can just hear the murmuring in the halls—"Rob Ryder, Hollywood Pimp." I've actually gotten a couple of emails from an ESPN insider—there are a bunch of guys there who absolutely hate my column. I mean, with a vengeance. There are two possible responses to detractors like this:

Response #1: Sorry, but what can I say? It's not for everyone.

Response #2: Fuck you.

Here's an update on some of the projects I'm pursuing.

GLORY ROAD

This is the Jerry Bruckheimer college hoops movie with Ben Affleck that's gonna start shooting this August. They've had my sports adviser résumé for three weeks.

Andy Given, the line producer, finally calls back. They're going with the guy who did the football in *Remember the Titans*. They're loyal to the people they've worked with. Fine by me; we need more loyalty in Hollywood. Besides, the idea of spending four months in El Paso takes the shine right off that one. I tell him that if they've got some heavy basketball days, I'll be available to come in, help direct traffic.

I just hope this other guy realizes that the way you can bark and scream at football players just won't work with basketball players. It's apples and anarchists.

A lot can go wrong on these expensive shooting days. An arena filled with extras and so much coverage to nail down in so few hours.

I once got called in on a movie when the director realized the two teams never switched baskets after the halftime of the climactic game. Yikes. That's a costly mistake.

Andy Given says, "By the way, have you been writing about us in some column?"

Whoops. Maybe there's another reason I'm not getting this job. What was that crack I made about Ben Affleck trading in his sleek Armani for a JC Penney plaid sport coat circa 1966?

"Uh, yeah," I answer. "On ESPN.com. You know, just having some fun describing the process."

"Yeah, someone told me, but when I went there, I couldn't find it."

"Yeah, no one can."

(That's one reason I'm so loosey-goosey about pissing people off. Since my "Hollywood Jock" column has been moved from the testosterone-laden Page 2 to the estrogen-intense Page 3, I've been pretty much lost in the ethernet.)

Andy and I wrap up amicably. I wish him well—*Glory Road* is a story worth telling. I remember to mention Saul Smith, the former Kentucky player whose father, Tubby, still coaches there. Maybe Saul can help out, act as liaison with the University of Kentucky.

Andy Given says, "Have him give me a call."

So Saul, if you're reading this, email me and I'll give you Andy's number. But you're better off not using my name.

ZULU *WAVE*

Last week George "Wildman" Walczak and I pitched *Zulu Wave* over at Paramount. A young black surfer in South Africa breaks the chains of apartheid.

It's a great idea, the pitch went great, but we still haven't heard back. Hey, that's cool. Given his recent track record, when John Goldwyn passes on your idea, you know you're doing something right.

Nothing personal, but in Hollywood, you gotta have that attitude: "What are you crazy!? You're not gonna produce our movie!? What are you out of your mind!?"

I'm serious. If you don't feel that way about your own stuff, how do you expect anyone else to?

Here's a tip, though: You can't fake it. If you try to manufacture a movie from the outside in, try to make it as formulaic and commercial as possible, all you're gonna end up with is a big pile of mush.

The trick is to start out with something you're passionate about—a dynamic situation, complex characters.

Then sell it to a studio and let them turn it into mush for you.

As George teaches at AFI: "Moviemaking is the slow disintegration of a good idea."

Am I being too harsh? Hollywood has given us some damn fine movies lately. But, hey, why let fairness get in the way of a cheap laugh?

Have you seen Eddie Murphy in *Haunted Mansion*? If not, don't.

Anyway, there's a producer in London who's chasing *Zulu Wave* hard. Says he's got some money for us. Plus I'm meeting tomorrow over at Fox

Television Pictures, with David Madden, who's expressing interest.

And we're going out to others. There's a deal there somewhere. The trick is to make it with the right people and get the damn thing made.

94 FEET

College hoops. All one game, told from both sides' POV. Shoot two endings—either team can win. It's hard-core. Using real players, young NBA guys, so we don't have to fake the ball.

You know how in most sports movies the producers say, "Yeah, it's a basketball movie, but it's really about a lot more than just basketball"?

Well, *94 Feet* is a movie about basketball. Just basketball. An inside look. And if that's not enough to hold your interest, you're what's known in the sports marketing world as a "casual fan."

And *94 Feet* ain't for you.

I call Baron Davis. He expressed interest some weeks ago, and now his season with the Hornets is over.

I tell him I've scheduled a meeting with Ice Cube's producing partner, Matt Alvarez. If Baron can help deliver players, Ice Cube and his company, Cube Vision, might just get on board.

Plus there's a great part for an assistant coach who ends up taking over when his boss is ejected. (But does he win the game in typical Hollywood fashion? Not necessarily; remember, two endings.)

The meeting at Cube Vision is loaded with possibilities. Ice Cube is icy hot right now, after the *Barbershop* and *Friday* movies and all the other stuff he's got going.

FOURTH AND LONG GONE

A college football comedy about a black assistant coach who takes over a floundering D1 program in the Deep South. Based on a very funny novel by the crusty old coach Pepper Rodgers.

The producer, John Carls, and I are working on the outline to run past Ron Shelton.

THE 4MAN LEAGUE

A few years ago, I experienced one of those crushing Hollywood disappointments that made me declare, "Okay, that's it! Fuck this town. I can't stand it anymore! I'm gonna find another way to make a living."

My wife tried to mask her joy and relief. "Do you mean it, honey?"

"Yeah, I'm done."

"So you're going to get a job? A steady job?"

"Me? What kind of man do you think I am? I'm gonna start a professional basketball league."

(I thank God every day that this woman and I are still married.)

So I assembled a terrific management team, we formed an LLC, raised over $300,000 of seed money, and we're still pounding away to get this thing launched.

But unlike minor league baseball, indoor football, and even hockey, minor league basketball has a dismal track record. Look at the CBA, the USBL, even the mighty NBA's NBDL. Disasters all.

Then there's the newly revived ABA, where the thinking seems to be, Hey, guys, let's see which team loses the most money the fastest.

I've got a friend who worked for a USBL team. They made a sponsorship deal with Adidas. Free shoes. But the players never got them. The owner started bootlegging them through the local shoe stores.

Minor league basketball is that kind of business.

But we've come up with a great concept: four-on-four full-court basketball. Fast-breaking, slam-dunking, fan-pleasing hoops.

We've run dozens of trial games right in UCLA's Pauley Pavilion.

We've messed with the rules to eliminate all those stoppages in the fourth quarter. We've even introduced a hockey-style power play to replace team penalty free throws.

Power-Play Basketball. The four-on-three for one possession is great fun, and a great marketing hook.

Plus, it's a game designed for the television viewer. And we'll be running during the summer TV doldrums.

It turns out my book agent, Matthew Guma, is a hoops fanatic. A Carolina Tarheel. He loves the idea and has joined our team.

We had a recent high-level conversation with a network executive. Then Guma brought me into Creative Artists Agency to talk sponsorship possibilities. There's clearly interest in our concept.

But here's the deal about TV. Specifically sports and TV.

With a few exceptions, the ratings generally suck.

And if you're not a major league, no network's gonna give you a dime for your games.

The best you can hope for is some kind of revenue share. The league covers all the costs (including TV production), the network provides the time, then you split the ad revenue. Even the National Hockey League (with its crappy ratings) was forced to turn to this formula.

You'd be shocked at how little revenue a 30 second commercial on a national cable network can generate. Five hundred dollars. Maybe a thousand.

And get this—you can buy a 30 second spot on a Fox Sports Net regional station for under a hundred bucks. Swear to God. Wanna sell that old three-speed in the basement? Forget eBay—run your own commercial on Fox Sports Net.

After much mashing, rehashing, and gnashing of teeth, our business model finally makes sense. But for the league to be successful, we need to be on television.

That's what will draw a crowd, attract better sponsors, add legitimacy. Ask the guys running Arena Football. They haven't shared a dime of ad revenue with NBC, but attendance is way up.

If you see it on television, it's gotta be real.

So that's my week. I'm almost out.

But speaking of real, I'm getting a lot of email inquiries about reality television.

Everybody and his mother seems to have an idea—these things are popping up like mushrooms in a cow pasture.

Ooops—there's another one; everybody and his mother—*Oedipus Erexus*, a reality show (hosted by Jerry Springer) that follows six college guys as they try to seduce their mothers without them knowing it. Or maybe I should make it mothers-in-law. Get to first base, you survive the first cut. Get to second . . . well, you get the idea. Hey, this isn't bad. And then you could reverse the process. Mothers-in-law trying to . . .

You know what? I'm tired (and you must be exhausted if you're even still with me). So all those great ideas you have? First off, register them at www.wga.org. Then save 'em till next week.

I leave you with this thought:

As long as our guys are fighting and dying on foreign soil, we are obligated to think of them every single day.

28

YEARNING FOR THE CORPORATE TEAT

Both my meetings got pushed. This happens all the time. No one's saying these executives do it on purpose, but it chops you down a notch. The message is clear—something more important than you came up.

Hollywood is filled with thousands and thousands of ambitious people, all jockeying for position, looking to score.

It's overwhelming at times—a giant pulsating nuclear mass of crazed quarks, free radicals, and jimmy neutrons, all bouncing off the walls, morphing, careening, trying not to fall off the edge, to get spun out into oblivion, or to collide with some equally unstable isotope.

Television, movies, music, dance, theater, a couple hundred thousand actors, singers, producers, writers, executives, managers, directors, agents . . .

The streets and buildings are vibrating with our (far too often) unrequited lust for money and recognition.

And in each special universe, a pecking order is established. These processes are as varied as they are ridiculous.

A Rolex watch. A seat at a table. An automobile. An unreturned phone call. A postponed meeting. A corner office. An office no bigger than a closet.

I once had a junior agent at William Morris . . . wait, how did that happen? I was on a roll, three deals in a row, but somehow my original agent was oh-so-subtly replaced by a greenhorn.

"Rob, this is J.L. He'll be working with us on your future deals."

My first demotion. And it wasn't even from a job. But you can't blame the system, you can't blame the agent. It's just the way it goes.

Ask Joe Eszterhas, once one of the highest-paid screenwriters in Hollywood (who was last seen stewing in some field in Ohio).

Anyway, J.L. and I didn't exactly tear up the town, and he was soon yearning for greener pastures. So he gracefully exited William Morris and went to work for the producer Edgar Scherick.

Big mistake.

His tenure was short; I don't know why. I went over to Scherick's to meet with him one day, and the receptionist showed me into a closet.

I'm not talking about a small office here. I'm not talking about some tiny, windowless cubicle.

I'm talking closet.

J.L. was sitting there on a ratty swivel chair. Was there room for a desk? I don't remember. There was barely room for me to sit in the folding chair across from him. I remember us bumping knees.

"So, J," I asked. "How's it going?"

"How do you think?"

Contractually, these situations can get sticky. If you quit, no severance. No golden parachute. Or silver in this case. Pewter . . . ? Lead . . . ?

So junior level executives like J.L. will often tough it out. Face one humiliation after another: "I won't quit. You have to fire me. You have to pay me to vacate the premises."

"Hey, J," I said, "I know it's tough, but if I were you, I wouldn't schedule meetings in this office."

"Yeah?"

"Yeah, it's uh . . . it's not big enough."

"Yeah."

I felt for him, I really did. I still do. Even though, years later, he told me over too many cocktails at the old Hollywood Athletic Club on Sunset that he purposely tried to submarine my career.

I remember peering at him through the booze and cigarette smoke (that's how long ago it was). Thinking, What?! Did you really say that?

I let it pass.

He ended up back at the William Morris Agency. In corporate affairs. They're like the NBA. They take care of their own. These cash cows have multiple teats. Prove yourself a loyal member and you can gum your way into retirement.

For writers in Hollywood, there is no NBA. There is no corporate position at the William Morris Agency. There is the brass ring. There's the abyss. Or worst of all, there is the plight of intermittent reinforcement.

Teach a desperate rat to keep bashing his head into a cinder-block wall and every ninth or 16th or 37th time reward him with a morsel. You think that rat's gonna look elsewhere for sustenance?

Not if he's a writer.

Writers, of course, are convinced we get treated the worst. I've heard the argument that it's all due to jealousy: "They need us. They can't do what we do, ergo, they're jealous and they treat us like shit."

I don't buy it. Writers get treated badly because we're so annoying. Look at it from the agents and executives' POV. Endless waves of neurotic whiners, looking to get paid to sit around and make a bunch of crap up.

It's enough to make you want to kiss it all good-bye and go off and start a summer professional basketball league.

But then another morsel presents itself.

The Pepper Rodgers football novel, *Fourth and Long Gone.*

This job, I can taste. It's right there. The producer who holds the rights wants me to do it. It's timely, with all the recruiting fiascoes taking place. And Pepper and I have established a nice vibe. He pops out from the East Coast now and then.

I'm at the Hamburger Hamlet on San Vicente, sitting across from the producer, John Carls, and Pepper Rodgers. Head coach at Kansas, UCLA, Georgia Tech—Pepper's the real deal. Plus he played quarterback for the Yellow Jackets and even kicked the winning field goal in the Orange Bowl back in 1952.

Now he works for Dan Snyder, who happens to own the Washington Redskins. Pepper's on the far side of 70. He obviously takes good care of himself. I ask him what it's like to work for Snyder. He considers, then softly drawls, "That man is a stern taskmaster. But he's fair."

Some years ago, Pepper wrote *Fourth and Long Gone.*

And Pepper Rodgers knows how to tell a story:

"It was 1971 and we were going up to play Stanford. I addressed my team. 'Men, these Stanford boys have it all over you. In a couple of years they'll be driving the fanciest cars, they'll have the most money, and they'll marry the prettiest girls.

"But there's one thing you can do Saturday afternoon to make things right—you can kick their asses. And for the rest of your lives, no one can ever take that away from you."

Pepper sips from his iced tea and continues, "So we went up there and did exactly that—59 to 13. We had 621 yards rushing. Not total—I'm talking on the ground. We pulverized them.

"At the press conference afterward I said to all these reporters, 'There are four things I love in this life. Fast backs and big linemen, beautiful women and good music. The fast backs and big linemen I reserve for the afternoons. The beautiful women and good music I hold for after six o'clock.

"Then I looked at my watch and said, 'Gentlemen, it's after six o'clock,' and I walked out the door."

Pepper's one of the good guys. But as with every college football coach, there's an edge there, a mental toughness.

"Football is a violent game," he tells me. "And that will never change, and that's all right."

29

DAVENING TO "DODGEBALL"

When you've got something going, when you're closing in on a deal, you don't want to jinx it. You don't want to overreach, get too greedy, but you don't want to sell yourself short either.

Zulu Wave is that pitch my pal George "Wildman" Walczak, Mark Rogers, and I have developed.

So far, we've pitched at Paramount, Fox Television Pictures, and to a hotshot young producer in London. It's the kind of movie you really want to write. The kind you want to see made.

That's the angle the English producer is playing with us. "I can get this picture made. I've got the contacts, I've got the resources."

"What about the money to write it?" we ask.

"There'll be money for you guys. Maybe not as much as you'll find in Hollywood. But you've got a much better shot at having this film made with us than with anyone over there."

He might be right. Especially if we get a typical Hollywood development deal. Again, for every movie that a studio makes, there are at least 25 in development that never see the light of day.

Those are long odds.

But what about television? There are some fine movies being made for TV. Plus there's an immediacy to the process, and the audience reach is enormous. It's an attractive alternative.

I'm at a breakfast meeting with David Madden of Fox Television Pictures. John O'Groats on Pico. O'Groats is an old joint, the kind of place that gives L.A. a sense of place. (This is a city so desperate for identity it put Bob's Big Boy on the Register of Historic Places.)

Demonstrating a masochistic streak, Madden has been reading my column, week in, week out.

He first learned of *ZuluWave* when everybody else did (that means you too, Mom). He popped me an email—"Sounds like a good story."

Hence breakfast.

David tells me he's shot in South Africa. "Don't laugh," he says. "But we made *Home Alone* 4 there."

I laugh.

"Cape Town doubling for Chicago. It's a great place to make a movie. Beautiful country, terrific film crews. Plus it's cheaper than cheap. It's cheaper than Canada by half."

That's saying a lot. So many American movies and TV shows are shot in Toronto and Vancouver simply to save a buck. Paraphrasing James Carville—"It's the exchange rate, stupid." It's the reason why *The Simpsons* is drawn in Korea. Why Yahoo's 800 number will be answered by a Sikh in New Delhi.

"And they'll really love you over there," continues Madden. "Hollywood making a movie in South Africa that's actually about South Africa."

Madden says he'd like to move forward with *Zulu Wave*. Bring it to TNT, HBO maybe. But first he wants to run it by his old producing partner. This guy's got a development fund that he can spend as he wants.

If he bites, we'll get paid to write the script. Then we can run with it as a feature, and if that doesn't pan out, our fallback is a movie made for television. It's the perfect scenario. Madden and I finish breakfast and part ways.

A week later, George, Mark, and I are still waiting to hear. Staying cool and calm. Or doing our best. But this kind of waiting can really wind you up tight.

George goes to the movies to relieve his stress. He'll sit through a four-hour Dogma 95 film about Hungarian house painters. He'll go to the cheesiest romantic comedies. He loves a good action popcorn movie.

He and his wife, Liz, are sitting in a crowded theater, waiting for *The Day After Tomorrow* to start. This is the kind of movie I'll tell my kid to go see and tell me about. Not George. He's gotta be there.

Like most long-suffering screenwriters' wives, Liz wants to know but doesn't want to know. She wants to see her husband's dream fulfilled. She wants to see the money in the bank. But she's learned to keep her distance.

So many disappointments, the reinforcement so intermittent. But *Zulu Wave* got to her. She's tracking the strategy, staying positive. She reaches over in the dark and touches George's arm. "I can't stop thinking about it. There's got to be a home for it in Hollywood."

George answers through clenched teeth, "I don't want to talk about it, all right? I don't want to talk about it."

"All right, all right. I just think, with all the movies they're making these days . . ."

"You don't understand," he hisses. "You don't get it."

George turns back to the screen, trying to hold it together.

Suddenly the trailer for *Dodgeball* comes on. George's eyes bulge, his blood pressure spiking sharply. He starts involuntarily rocking back and forth. *Dodgeball*'s tag line—"Grab Life by the Ball."

Fifteen seconds into the trailer, and George is rocking like a maniac. Davening to *Dodgeball*, the theater screen his own personal Wailing Wall.

"Would you stop doing that, please?" It's the guy in front of him.

George is a Polack from Jersey City. "Go fuck yourself," he says.

"I'm uh . . . I'm calling the manager."

"Good, go call the manager. Go call the cops."

"I will. I will."

"Good! Do it! Call the FBI! Have me arrested. Put me out of my goddamn misery."

It's the hurt in George's eyes. The guy suddenly softens.

"What is *wrong* with you?"

"What? I'm tense, okay! I got a lotta stress in my life right now. I really do."

"Yeah?"

"Yeah. So I rock, okay? I'm rockin' my way through *Dodgeball* here. I'll be done in a minute, all right?"

"Yeah, okay. All right."

The trailer ends, the rocking slowly subsides.

"You better now?" the guy asks.

"Yeah, thanks."

"Enjoy the movie." The guy rolls his eyes at Liz and turns away.

Me, I try to channel my stress. Waiting for word on one deal, I plug away on others.

I spend some time on my book proposal. Set a couple meetings on this college football project, *Fourth and Long Gone.*

Then I get a call from Jason Jepson. He's a young public relations guy who's starting his own company, Screaming Monkey. He can put me together with the Maloof brothers. They own the Sacramento Kings. This I like. They're starting a movie production company.

My former teammate, Geoff Petrie, happens to be president of the Kings. Doing a great job, although they've been snakebit with injuries the last few seasons.

And our old coach at Princeton, Pete Carril, is on their bench playing Yoda. Not that the Carril connection's gonna help me. Petey and I didn't exactly see eye to eye. But anyone who knows Carril knows you can have your differences with the man and still be all right.

Jepson will try to set a meeting.

The next day, I get over to Revolution Studios to meet with Ice Cube's producing partner, Matt Alvarez, and his lieutenant, John Hayes. The meeting was set by Matthew Guma, my book agent, since my Hollywood agent, Jon Klane, has been acting more like a producer lately. Or a rock star.

The guys at Cube-Vision are smart, low-key. The company is riding a wave. They want to make smart choices now. We do the chitchat, getting-to-know-you, thing. We laugh and groan about *Soul Plane.*

John Hayes is African-American and *Soul Plane* is not his kind of movie. "If it does well, you can just imagine all the knockoffs we'll be seeing."

Soul Boat. Soul Bus. Soul Biscuit.

We talk about a couple things we might work on together. I'd like to develop *Hollywood Jock* the movie with them. The sensibility feels right. And it's great to get that African-American perspective.

Alvarez says there's nothing worse than a white screenwriter trying to write black dialogue. "It doesn't work. You're better off writing it straight and letting the actor bring the color to it."

I tell them I try to do readings of all the screenplays I'm working on, especially when there are minority characters. They can give you fresh language, the license you need, the stamp of approval. Plus it's a defense against getting mau-maued by Spike Lee.

I wonder what Spike thinks of Snoop . . . hmmm . . .

These table readings get you out of the house. They bring you closer to the process, actually giving you a chance to do some directing. It's great fun, and sometimes horribly revealing:

"So that scene, that's not very funny, huh?"

"No," an actor answers. "Not even close."

Others nod in grim agreement.

"So what does it need?"

"It needs to not be in the screenplay."

"Oh."

I'm not in a rush with Cube-Vision. We'll bring things along.

I get home to find an email from my editor at ESPN.com. "Will you be writing about the NBA playoffs?"

I watched game seven of the Eastern Conference Finals last week. Final score—69 to 65.

I email back, "No."

30

"MY KNIFE IS BIGGER THAN YOUR KNIFE"

Email responses to my columns have been coming in from soldiers, from Marines, from veterans. Thanking me for remembering them. Encouraging me to keep it up. And I sit here thinking, You're thanking me? Jesus Christ . . .

Our parents and grandparents got it right. In World War II, you couldn't get away from the war. Turn to the local sports pages, there'd be a remembrance of a Nebraska high school wrestler who met his match on Omaha Beach. Go to the movies, there'd be newsreels keeping the war front and center. Go to a Chicago White Sox game—volunteers collecting for war widows.

People sacrificed. They bought bonds, saved scrap metal, prayed, paid attention. All of them. A collective effort.

I just heard from a guy whose kid sister is in Iraq right now, serving as an army interpreter. This guy is sick with worry, but his sister's emails are filled with chatter about a potential boyfriend—a corporal she met in Baghdad.

Many of these troops are in their early and mid-twenties.

Then there's a young Texan, a vet from the first Gulf War, who writes how fresh out of Catholic high school he saw more violence in three days than we can imagine—searching, scrambling, trying to destroy a downed Harrier before the Iraqis found it. The coordinates all fucked up, Saddam's bad guys materializing out of nowhere. He wants to know if there's a movie in this. I don't know. His email is filled with some of the most beautiful, raw language I've read in years. The truth. I encourage him to write down the whole story, just the way it happened.

But how many *Black Hawk Downs* can we handle? Because from my seat in the bleachers, that's the true nature of war. It's a mess. It's ugly and gut-wrenching and nothing happens the way it's supposed to.

The Green Berets with John Wayne was one of the first movies produced about Vietnam. It ended with John Wayne standing triumphant on the beach at sunset.

The sun was setting in the east.

Then there are the three college guys who email about their fraternity brother in Iraq: "He just got extended again—running around in the desert chasing lunatics. We're rooting for him to get home."

This isn't about politics here. This is about doing right for those people who are doing right by us.

It's a hard, harsh world. It always has been.

Maybe this stuff is getting to me because of my boys. They're twelve and six. What does the future hold for them? And it's not just this latest war I'm talking about. It's the human condition. How do you teach them to navigate through this dangerous landscape called life?

I've got a buddy, Nick Stein, who puts it like this: "Look, here's life—we all start out lined up at the edge of a minefield, the gun goes off, and you start running. Suddenly BOOM, the guy alongside you blows up and you're like, Oh, jeez, but you just keep running. What else can you do? And it's like BOOM, BOOM, BOOM, and some of us, for whatever reason, get to keep running, and then we're old and that's the finish line. That's what life is."

But for me, you know what? That's not good enough.

There's strategy involved in life. There are secrets of navigation.

My kids are learning some of these secrets through sports.

Fittingly, it ain't all pretty.

Sure, there's the winning and losing. Most kids get that. We all do. But then there's the nasty stuff. I keep returning to abusive coaches. I'm witnessing too much of this stuff right now in my son Cole's Pony League.

It permeates the culture of sports.

Even a guy like Phil Jackson, Mr. Zen Master himself, has a nasty streak.

From the *L.A. Times*: "Phil Jackson enlists a willing young player to be a 'whipping boy' at practice, so he can yell without offending his big stars. He calls it a 'cheap practice' that many coaches employ. Horace Grant and Toni Kukoc filled the role in Chicago. Devean George and Slava Medvedenko now take the brunt of Jackson's instruction in L.A."

"A willing young player?" Yeah, right. "Hey, Coach, need someone to dump on? I'm your guy."

At least Phil Jackson is honest enough to admit it's a cheap practice.

Some coaches play it differently. Pete Carril, at Princeton, was an equal opportunity inflictor. Carril was a brilliant strategist, but not exactly a touchy-feely type.

On the team bus, we'd speculate about who'd be in his doghouse next. Singing lyrics from that old Blood, Sweat and Tears song: "What goes up, must come down, spinnin' wheel, got to go round."

But at least he spread the pain around.

It's worse when a coach focuses all his criticism on one or two guys. It creates an unhealthy dynamic—turning the other players into enablers. They witness the abuse, they know it's wrong, but hey, they're not on the receiving end, so they keep their mouths shut.

In World War II, these people were called Good Germans.

Am I stretching the analogy? Of course.

But that's what happens when you see way too many of these knuckle-scrapers laying this crap on our children. (I'm talking youth sports coaches here, guys pretending they're Bill Parcells.)

But that's no four-million-dollar professional athlete they're screaming at, it's an anxious nine-year-old trying to find his way.

My friend Elizabeth and I recently talked about just how intimidating, just how frightening men can be.

When it comes to men, two things are clear. One, physically, men are bigger, stronger, and hairy in all the wrong places. Two, we've got just about the most dangerous drug known to mankind coursing through our veins—it's called testosterone.

And here's a bright idea—on top of all these built-in aggressive tendencies men have (I'm talking xenophobia, homophobia, and the genetic need to dominate), how about we all get liquored up on Saturday nights and really embrace the beast within?

Sometimes I thank God I'm not raising girls.

Walter Hill, the director who gave us *Hard Times, The Warriors, 48 Hrs.*, and *The Long Riders*, has a penchant for casting scary men.

A penchant I saw played out in *Southern Comfort*.

It was 1981. I'd been in Hollywood about three years. Made a few bucks with a couple low-rent screenplay deals, but the money dried up and I had to take a real job.

I found myself in a huge empty warehouse down near LAX with my pal Joe Niesi, up on a scaffolding, painting endless rows of one-by-four cross-braces that were keeping the rafters from twisting.

I'd put in an eight-hour day, grind my way home up the 405 and across the 10, grab some Mexican takeout, and write till two, three, in the morning. Crash, get up, and do it all over again.

No shoes, no shirt, no life.

I was losing it. I was hurting. Then one evening I got home and there was a message on my machine from Walter Hill.

He had a part for me in his new movie, *Southern Comfort*.

Next thing I knew, I was sitting in first class on a flight to Shreveport, Louisiana. Walter had given me a stunt player role—a nameless bad guy Cajun. No dialogue, but a guaranteed three weeks at $1,750 per, plus stunt adjustments.

We'd spend the time running around the swamps of East Texas, shooting long-barreled rifles at Keith Carradine, Powers Boothe, and other members of their lost army reservist platoon.

Of course, I looked about as Cajun as Adam Sandler looks Norwegian, but hey, it's the movies. Walter knew I was broke. We'd been in touch since *The Warriors,* when he'd given me my first break. *Southern Comfort* was his latest bone for a hungry dog.

I'd been told there were three other guys flying in with me. I looked around first class. They weren't hard to spot.

They were the three guys who, traveling these days, would never get past security: Allan Graf, Ned Dowd. and Sonny Landham.

Graf was a huge man. He'd played offensive line at USC and had a Rose Bowl ring to prove it. He's gone on to be a football adviser and stunt coordinator on some big-time movies.

Ned Dowd was grizzled and hardened from years of minor league hockey. His sister, Nancy Dowd, wrote *Slap Shot* based on Ned's stories. Ned's worked his way up through the production to become a topflight producer and movie executive.

And then there was Sonny Landham. Sonny stood about six foot four and weighed a good 225—all muscle. Part Native American, part mongrel. Black straight hair, piercing black eyes. His face had that glazed, reconstructed look like it'd been through a few windshields, or seen the wrong end of a broken bottle.

We introduced ourselves, and this being first class, got drunk.

The plane landed in Dallas. There'd be a 30 minute wait before it headed on to Shreveport. Sonny got off the plane. "I need me a new pair of boots."

We arrived in Shreveport and didn't see Sonny for three days. The production manager was pissed, but it didn't matter because all we did was sit around the hotel, trying to stay out of trouble.

Shreveport's the kind of town, when you're at the hotel bar and a young woman opens her purse, you're just as likely to spot a Smith and Wesson as a pack of Marlboros.

We were on call. Making decent money to do nothing.

Finally, Sonny arrived.

I was hanging in my room when the phone rang. Sonny was inviting all us "Cajuns" up to his room for a drink. Plus he had something to show us, he said.

I walked down the hall, climbed a flight of stairs, walked down another hall, and finally knocked on his door.

"Come on in!" I heard him yell.

I swung open the door.

Sonny was standing there, wearing his new snakeskin cowboy boots. And nothing else. The man was stark naked. Plus he was holding an extremely large and very sharp hunting knife in his right hand. The exact same knife he wielded in 48 Hrs.

This is called making a statement.

"Hey," he said with a maniacal grin. "What do you think of this knife?"

"I, uh . . . That's a pretty big knife."

"You got that straight," he said.

I pondered my predicament. They didn't teach you about this shit growing up in the suburbs of New Jersey.

"Come on in," said Sonny. "Have a drink."

"You know what?" I said. "I think I hear my phone ringing." And I bolted down the hallway.

Like Elizabeth says, "Men can be very scary."

Sonny Landham turned out to be a stand-up guy. He'd done three years in the army before becoming an actor. He was a real pro in front of the cameras—slogging through the freezing swamps with the rest of us, soaking wet and miserable. No bullshit. No whining.

In sports, in the movies, in life, you find a way to deal with these guys.

Hey, you got a bigger knife than me, that's fine. I'll grant you that. And in the meantime, we can have a laugh, do a good job together, make a few bucks, and everybody goes home in one piece.

You gotta navigate your way. Finesse, massage, outwit, take a stand, acquiesce. In a world filled with dangerous men—men way more desperate than you—you've got to choose your fights wisely.

You've got to make allies of potential adversaries.

Sonny Landham is the kind of guy you want to have your back in a streetfight, be it in East Saint Louis or Kandahar, Afghanistan, the 38th Parallel or Iwo Jima.

Pants or no pants.

But it's one thing to tell a story like this in a bar and another to put it in a column. My wife winced as she read it. "Was that such a good idea to write about that guy like that?"

"What—Sonny? I don't know. Why not? I gotta be outrageous, right?"

"You don't think he's gonna be angry?"

"What, that he was naked?"

"Yeah. And holding a knife."

"Nah, it's fine. He'll get a kick out of it."

But I'm suddenly filled with dread. Late that night, I Google "Sonny Landham," only to discover that he's running for governor of Tennessee. Swear to God. Here's a blurb I find on him from a blog called polstate.com—

> The first of the would-be Governors is Independent candidate William "Sonny" Landham. Sonny had a pretty good career as a character actor in Hollywood. He started by working in about 9 porn movies including such forgettable titles as *Naked Came the Stranger* and *Slippery When Wet*. He later had a lot of minor tough guy roles in such movies as *Predator* (with Arnold Schwarzenegger), *Lock Up* (with Sylvester Stallone), and *48 Hrs.* (with Eddie Murphy and Nick Nolte). Sonny was originally planning to run as a Republican but changed to Independent when it became obvious that the Republican ticket was getting crowded. Sonny's issues revolve around children. Specifically he thinks that the courts have way too much sway in divorce and child custody proceedings. This stand may stem from a particularly messy divorce he

> had from a former wife. She accused him of stalking her and he ended up serving almost 3 years in prison in the 90s before the conviction was overturned.

I wake my wife to assure her there's nothing to worry about. He served only three years, and he wasn't even guilty!

31

"DON'T SMELL 'EM, SELL 'EM."

The days are taking on a crazed feeling. So many projects to juggle, so much information to process. And now my Hollywood agent, Jon Klane, has decided to turn himself into a producer, leaving me hanging.

It's a shame. Not that he's been doing that much for me (he hates sports), but he's one of the coolest guys in town. A writer client went to him with a problem. "Get yourself a lawyer," Klane said. "I'm a huckster, not a shyster."

My book guy, Matthew Guma, has been great, but he's back in New York. So I'm scrambling, partnering up with sundry bedfellows—standing outside the studios, throwing our shit against the wall like thousands of others, seeing what's gonna stick.

These two young English producers are coming on strong about our South African surfing story, *Zulu Wave*. They offer a compelling argument—it stands a much greater chance of getting made through them as an indie feature with international financing.

Bend It Like Beckham. *Whale Rider*. *Chariots of Fire*. That's the sensibility of *Zulu Wave*.

I call an old friend, Norman Stephens, for advice. Norman was my first literary agent. I remember him repeating what an old-timer told him years ago—"Don't smell 'em, sell 'em."

Norman soon left agenting (was it me?) and became an executive, and then a producer—specializing in television movies and miniseries. He's had an impressive run in a very difficult field. I call his cell—never knowing whether I'll reach him in Hollywood, Bulgaria, Ireland, Australia, or Toronto.

(The Civil War movie *Cold Mountain* was shot in Romania, by the way.)

Norman answers. Miraculously, he's in L.A. He tells me he's producing *The Last Days of Pompeii* for A&E, and it's been a struggle.

They hired a talented writer, Benedict Fitzgerald, who turned in a credible first draft. But Fitzgerald also wrote *The Passion of the Christ*, and that became a problem. See, Mel Gibson paid him Writers Guild minimum to write the screenplay, but gave him three gross profit participation points in the movie—the points defined exactly as Mel Gibson's were.

Those three points are conservatively estimated to be worth between 10 and 15 million dollars to this once struggling screenwriter.

The money became a bit of a distraction (imagine that). So they finally brought on another writer, but it looks like they're losing the summer. (Unless, of course, they head south below the equator.)

Norman listens to our dilemma on *Zulu Wave*. He recommends forgoing Hollywood, pursuing the international route.

"The most successful of these movies have a homegrown feel. Look for European money. Follow up with these English producers. Find a hot young South African director. I'll put you in touch with a couple of people."

"Thanks, Norman."

"Send me the treatment," he continues. "And there's something you can do for me. I've got a friend who's got the documentary rights to the story of the Syracuse championship lacrosse team in the 1950s. The team with Jim Brown. Frank Deford is involved. They're looking for a sports documentary producer. Know anyone?"

Sure I know someone. For a project like that . . . First of all, Frank Deford's one of my favorite writers.

And Jim Brown I consider a hero.

Not enough people know that Jim Brown was just about the most ferocious lacrosse player in NCAA history. First, imagine Jim Brown—all six foot two, 230 pounds of him—coming straight at you through a hole in the offensive line with a football in his hands.

Now imagine that same Jim Brown barreling down on you, swinging a 3-foot-long, hardwood lacrosse stick.

This man is not a saint. He's done some ugly and violent things in his time. But when it comes to social justice, where so many talk the talk, Jim Brown walks the walk.

He knows the trouble brewing in our streets and in our prisons. Trouble for all of us. He's been shouting about it, putting in some serious time on it. Is anybody listening?

The Syracuse story would make for a compelling documentary. I put Norman in touch with Brett Rapkin and the award-winning Black Canyon Productions.

There's lots of great stuff happening in the sports documentary world. It'd be great to get lost in old tapes and new possibilities. But I've got work to do.

I get off the phone and consult today's three-by-five to-do list.

The fourth item reads, Answer emails.

The emails I've been receiving in response to the column have been thoughtful, poignant, heart-wrenching, absurd, hilarious, and sometimes just too cool.

A guy named Gilbert Kamuntu writes from Kampala, Uganda, that he's been reading my column since I debuted last November and wants me to succeed in all things.

Here's how he ends:

> "And please, to show you and any potential book publishers how international your column is, could you very kindly give a shout to my friends (also great fans of yours), Naaku Paul Birabi from Lagos, Nigeria and Frederick Gikonyo Kambo from Nairobi,

> Kenya. And also to my friend Edward Kibuuka here in Kampala who thinks that Allen Iverson should be in the Hall-of-Fame."

So yeah, here's a shout for my new friends in Africa.

I appreciate what Gilbert said about book publishers. But it's crazy, me being allowed to pimp my own projects like this in my column. And it may be backfiring.

I had a falling-out with a great old friend and wonderful writer, Robert Ward, who thought I was overdoing it. "You'd put Sammy Glick to shame," he wrote in an email. (Never say anything nasty in an email, I once heard it said.)

I iced him for a few weeks. Passive-aggression. The power of silence. But we've patched it up. The friendship is too important. Besides, he was right. I was overdoing it.

I'm trying to be respectful here. There aren't exactly rules for this kind of thing. You gotta pay attention when talented people criticize. And that goes for you too, Sports Guy. Next time I mention *The Warriors*, I'll first give you a heads-up.

But in the words of Missy Elliott: "Is it worth it, let me work it. I put my thing down, flip it and reverse it."

And if I'm sounding fractured, it's the nature of the beast. My editor's been requesting that I make my columns more seamless. I keep saying, you want seamless, you're talking to the wrong guy. You want seamless, try panty hose.

I suggest that we just draw lines across the page when I finish one bit and lurch into another. I see this all the time—columns made up of totally unrelated items, separated by lines. What's the big deal?

Imagine a thick black line right about here.

Julian "J.J." Jones has the young man's dream job, bar none.

Forget Donald Trump's apprentice. Forget Hugh Hefner's pool guy. Forget ESPN's rookie coanchor.

J.J. Jones is Gavin Maloof's right-hand man. He's his gofer, his aide-de-camp. His special envoy, his go-to guy, his eyes and ears for all things hip.

For those of you living in Siberia (and I know you're out there), Gavin Maloof and his older brother Joe and the rest of the Maloofs own the Palms Hotel in Las Vegas and the Sacramento Kings. Plus they own new music, television and movie production companies in Los Angeles and all sorts of other stuff.

I got hooked up with J.J. through this hotshot young PR guy Jason Jepson. They once worked together at the hugely successful Internet Home Loan Center in Orange County.

Julian Jones is 22 years old, an entrepreneur and a rapper. I get him on the phone and ask him his story. In a torrent of words, J.J. lays out the broad strokes—he was born in L.A., his parents divorced and he moved with his mom to Orange County, but his dad stayed behind in Inglewood and as J.J. puts it, he got the best of both worlds.

"Got my book learning in Orange County, got my street smarts back home with my dad."

J.J. tells me he's always been hustling to make a buck, to make his mark. He was selling CDs out of a car trunk when he was 14. Writing music, organizing parties. After high school, J.J. did some college-specific courses like international marketing and business law, but he got antsy and jumped into the private sector. Selling loans by day, working parties by night.

Then, one weekend, he was partying in Vegas at the Palms Hotel and he met Gavin Maloof. He's worked for him ever since.

"I do everything that needs to get done," he says in a rush. "I go to the bank, I schedule meetings, I order room service, I listen to new demos. I can tell when a song's a hit, and I'm not afraid to speak my mind. I think that's why these guys hired me. But don't get me wrong, I'll go pick up the laundry, you know what I'm saying? I'm a young guy. I'll do what it takes. But I know what's hot and I believe that's my true value to the Maloof organization."

J.J. splits his time between Las Vegas, Los Angeles, and Sacramento. It's a dynamic but exhausting schedule. "These guys work hard," he tells me. "Eighty percent of their travel they do because they got to, not 'cause they want to."

The Maloofs, as businessmen, live by a creed—the customer comes first.

J.J. and I talk about how important it is for young guys on the rise not to step on too many toes. It's a delicate situation—you want your voice to be heard without pissing people off.

Then J.J. inquires about me. I lay out *94 Feet*, the college basketball script, and *Hoop deVille*, the music/basketball show that will play great in Vegas. I tell him Bob McPhee's been helping me along—he's a theater producer on a nice roll (*Avenue Q* is among several of the shows he's been involved in). J.J. gives me the number for Maloof TV in Santa Monica and says they'll be expecting my call.

We wrap up with him telling me that Gavin Maloof is so cool, he's letting him keep his rapper dream alive, giving him the time to head into the recording studio this week with the all-time gangster music producer, Suge Knight. (And I think, Yeah, either that, or he's trying to get rid of you. "Hey, Suge, that's just a joke, right? Kinda like Weird Al Yankovic and Coolio, huh? You know, 'Amish Paradise'? No harm no foul . . . Suge . . . Suge?!?!")

Joking aside, I hang up, impressed. Knowing that ambitious young people like Julian Jones make the world go round. Remembering J.J.'s parting words:

"My mom always said, 'Closed mouths don't get fed.'"

32

BEST FOREIGN SPORTS MOVIES

ESPN is airing a show tonight about the best sports movies of the last 25 years. The folks in Bristol requested that this week's column tie into that somehow.

What?! You mean stop torturing everyone with my whiny, self-serving tales of Hollywood and write about something else for a change?

No problem. (But clearly, something is amiss at ESPN.com. Methinks they're about to pulleth the plugeth. So what do I do instead of getting some straight answers about what's working and what's not? I decide to push the envelope.) So here's this week's column:

When I reviewed the list of the top 25 sports movies on ESPN.com, I just couldn't think of anything to say that hadn't been said before.

Not that there weren't some fine movies there. Great stuff. Crisp writing, gutsy performances. Whimsy, heartbreak, you name it.

But I simply couldn't come up with a fresh take. Until finally it hit me. There were only two movies on it that hadn't been made in America by Americans.

Aha. What about all the great FOREIGN sports movies? Why not write about them?

But that, too, was easier said than done. I could barely conjure up five before my already meager supply of brain cells started to sizzle and pop.

So I decided to turn to the trusty readers of my column.

Actually, to those readers in my email address book. See, everyone who's written me over the last several months has been logged in.

I figured, Why not just do a mass emailing. Take a poll. If there's one thing I've learned about my readership, it's this: They're fiercely intelligent, well-reasoned, worldly and wise.

So I mailed it off: "What are your favorite FOREIGN sports movies?"

And here are some of their replies:

"Who gives a shit?"

"Is there even such a thing as a good foreign movie? Good luck with the horrible assignment."

"Does Iowa count? If so, *Field of Dreams*."

"Can we please consider *Rookie of the Year* a foreign movie so the world can't hold that against us any longer? Man, that one sucked."

"From here in New Zealand, *Slap Shot* and *The Longest Yard* look pretty foreign, dude. I love you Yanks to death, man, but I wish you'd collectively work out that 'foreign' is all about perspective."

"*The Boxer* . . . Jim Sheridan, Daniel Day-Lewis. They did amazing things with color and silence in this movie. Plus, some of the most realistic 'movie boxing' I've seen."

"*Bend It Like Beckham* was great."

(And a big vote-getter.)

"Does *Rumble in the Jungle* count?"

(That'd be *When We Were Kings*, by Leon Gast, and sure, why not? A highly entertaining documentary film about the Ali-Foreman fight in Zaire.)

"Remake of *The Longest Yard* a few years back using soccer in place of football. It was called *The Mean Machine*."

"*Cool Runnings*."

"*The Loneliness of the Long Distance Runner*, in which a young Tom Courtenay running for a prison team refuses to win as an act of courage. The anti-Rocky ending, perfect."

"*The Goalie's Anxiety at the Penalty Kick*."

(One of my favorites, just for the title alone. But you want bleak? Here's what IMDb.com had to say—"A goalkeeper is ejected during a game for foul play. He leaves the field and goes to spend the night with a cinema cashier. He then proceeds to strangle her the morning after." Don't expect a remake from Disney anytime soon.)

"*The Seventh Seal*. Okay, chess isn't a sport, but it's really cool when he plays chess with death."

"*This Sporting Life*, with Richard Harris and Rachel Roberts. Directed by Lindsay Anderson. One of those British "kitchen sink" films of the early 1960s about a rugby player in northern England."

(This is a favorite of writer-director Ron Shelton, who happens to have directed four of the top 25 listed at ESPN. Ron also logged in with *The Loneliness of the Long Distance Runner* and *Touching the Void*, a British movie about two mountain climbers who get stuck, and one has to cut the rope, letting the other fall.)

(Sounds a lot like Hollywood.)

"It doesn't get any better than *Shaolin Soccer*. Not that I have actually seen it. . . . I can just tell."

(*Shaolin Soccer* received many votes. People say it's hilarious, but don't get the dubbed version.)

"How about *Men with Brooms?* That is, if you consider Canada foreign."

"*Iron Ladies* from Thailand, about ladyboy volleyball. Also from Thailand is *Beautiful Boxer*, about a transvestite muay thai kickboxer. Thais love their sports and their ladyboys; go figure."

"*Olympia*, by Leni Riefenstahl." (Documentary about the Berlin Olympics in 1936. Compelling stuff.)

"*The Great Race*." (Uh, okay.)

"*Gallipoli* is one of my favorites. It stars a very young Mel Gibson as a competitive Aussie runner who follows a friend in enlisting for service during WWI and witnesses the perils and futility of war."

"I'll never forget a trip to the cinema in London to see *Billy the Kid and the Green Baise Vampire*, a musical billiards showdown."

(Somehow, I don't think you'll find that one in Blockbuster.)

"What's that movie where the guy that played Goose is a paintball player in school, and he gets caught up in international spy games during the cold war in East Germany. Does paintball count as a sport? I don't know, I just wanted an excuse to see that Natasha chick's picture again." (*Gotcha!*)

"*Chariots of Fire* for obvious reasons. It's a movie you can enjoy even if you don't know what a fartlek is."

(Thank God for that. *Chariots of Fire* was the overall favorite, by the way, with *Victory* coming in a close second.)

"*Bad News Bears Go to Japan.*"

"There is a French-Canadian film about hockey named *Les Boys* which is actually pretty funny, even for the French."

"*Lord of the Flies* :-)"

"*The Power of One*. Boxing, 1950s South Africa, Stephen Dorff/Morgan Freeman. Pretty good stuff."

"I believe *Victory* was an international production. If so, that would rank # 1 with me!!! Pele, soccer, whipping some Aryan arse . . . can't beat that with a baseball bat!!!!"

(Many many votes for *Victory*. It's a bit of a stretch, since it was directed by John Huston, produced by Freddie Fields, and starred Sylvester Stallone, but why not? It's worth viewing.)

"If your definition of sport is broad enough to include 'dancesport,' then put me down for the Aussie comedy *Strictly Ballroom*, where the competition is far more cutthroat than it ever gets in those sissy American sports of baseball, football, and basketball."

(*Strictly Ballroom* received three other votes. Hey, guys, you want me to put you in touch with each other?)

"If Indiana is considered a foreign country, then *Hoosiers*."

"*The Ten Commandments*. Final score: Israelites 10, Egyptians 0."

(And last but not least:)

"*Karate Kid Goes to Hawaii*? Is that foreign?"

(Okay, I'm out. And again, say a prayer for our troops, who have their own foreign movie going on.)

33

"OUR MODELS WILL BE NAKED"

Okay, I'm closing in on a couple of deals now, and it's time to start playing my cards closer to the vest. I'm learning the hard way that people don't necessarily want their business discussed in public.

I've crossed the line a couple of times—wasn't clear when something was on or off the record. And maybe I've been a bit too critical. Hey, my bad, okay? No harm no foul, right? Oh, yeah? So take me out back and shoot me.

I probably got off on the wrong foot months ago when I trashed *Eddie* in my column and the two producers called me separately not to object, but to share in the laugh.

David Permut was even-keeled about it: "Look, not every movie works out. You've got to keep moving, that's all."

Mark Burg didn't mind me writing about how he'd gotten reamed by Whoopi. "If you don't have a thick skin in this town, you don't have a chance."

So early on, I dodged a bullet and was deluded into thinking, Hey, these guys don't mind some good-natured ribbing. All publicity is good publicity, right?

Plus it's exhilarating being a loose cannon.

You know those great news stories where a guy steals a tank or a bulldozer and goes on a rampage? It's like that. At least until someone gets really pissed off and tries to spike your barrel. Ouch.

No names, please. No names. This is a dangerous environment we're living in.

Mike Ovitz was a ruthless uber-agent. You know what he and Saddam Hussein have in common? Even though they've both been rendered powerless, there's still this nagging fear that they'll come back and get you somehow if you take a swipe at them.

On the other hand, playing the upstart, criticizing the powers that be, has been a proven path to success.

Consider Joe Namath and Super Bowl III.

And consider François Truffaut.

(For those of you wondering if Truffaut was that left-winger who played for the Maple Leafs back in the '60s, you're wrong. That would be Jean-Luc Godard.)

Truffaut was a young film critic who absolutely blasted the most revered French directors of his day and was quickly offered the opportunity to direct something himself.

So I'm thinking, Hey, since it worked for him . . .

But then again, Truffaut had *400 Blows* up his sleeve—a beautiful, haunting story that I strongly suggest you hunt down the next time you're in the video store, eyes glazed over like a couple of Krispy Kreme doughnut holes.

And as long as I'm rambling here, you owe it to yourself to see *The Station Agent*. Now I can really hear the groans. Look, I know it's a movie about a train-obsessed dwarf, a grieving mother, and a motor-mouthed Puerto Rican, but at least it's in English.

It was made in New Jersey, for God's sake. So see it. If you're still reading this book, you'll enjoy this movie. I guarantee it.

And if you're looking for a good, stupid rental, try *Dodgeball*.

My screenwriting pal George was dragged along by his two boys, Dash and Tristan, and you know what? They laughed.

George's report: "Stupid, but surprisingly funny, with several cool reversals and a wonderful performance by Vince Vaughn."

Yes, this is the same George who went into paroxysms of pain when he first saw the *Dodgeball* trailer and thought, This is the kind of junk Hollywood wants to make.

But he was wrong. And so was I.

"In Hollywood, nobody knows anything."

Just ask the company that spent over 100 million dollars on the remake of *Around the World in 80 Days* (domestic gross to date: barely 20 million).

And when it comes to giving *Dodgeball* its props, never, ever underestimate the value of a cheap laugh. They don't come easy. Just asks the guys who made that recent Bruce Willis clunker, *The Whole Ten Yards.*

As long as we've been trotting out the clichés here, how about this one:

"No one ever sets out to make a bad movie."

It's just that there are so many things that can go wrong.

And for all you right-side thinkers out there (or is it left-side?), I'm sorry but it's not like building a building. There's magic involved. And where you find magic, you find bad magic, and bad magic's got a mind of its own.

There's nothing worse than being on a snakebit movie that still has weeks left to shoot, and you realize, God, there's no saving this soggy sack of celluloid. Can't we just stick a fork in it and get on to the next one?

Sorry.

So you might ask, Why do people even bother?

Because when you get it right, whether you're the director, writer, cameraman, bit player, grip, assistant editor, or just another guy who bought a ticket, there's nothing like a good movie.

That's why people stay in this game.

That and the money. Plus Los Angeles has a better climate than Las Vegas.

Stay in the game. Stay in the game.

It was Woody Allen who said that 90 percent of success is just showing up. But that's particularly tough for unemployed screenwriters.

A couple months back I went over to Sony to visit with a writer-producer I used to hoop with, Alex Siskin. Siskin contends that writers can no longer simply create the work and let their agents do the selling. Every screenwriter in Hollywood should start thinking of himself as a producer as well.

And I thought, Great, Alex, you just quadrupled the number of unemployed producers in this town.

But he's right. You've got to push your own work. It's not enough to stay holed up in your little cubby being brilliant.

Last week, Siskin emailed me his entry for "Best FOREIGN Sports Movie": *Bend It Like Beckham*. But it seems he had a vested interest in this mention. He and his partner over at Sony have been working for years to get the movie version of *I Dream of Jeannie* off the ground.

(If that just made you wince, don't forget, "Nobody knows anything.")

And these guys are being smart by bringing on Gurinder Chadha, the director of *Bend It Like Beckham*.

It's not the obvious choice. She brings a fresh eye and a nice way with actors, and if they have a good script, who knows, it could be a big hit.

So thanks, Alex, for your contribution last week, and good luck with *Jeannie*.

And thank you, dear readers, for making last week's column on best foreign sports movies such a resounding success. (At least for me—I didn't have to do much writing, and there were a lot of cheap laughs.)

Did anyone catch ESPN's "Best Sports Movies of the Last 25 Years"?

That was pretty cool. And I liked the choice of *Hoosiers* as number one, even if I didn't work on it.

But when I saw all those talking heads being interviewed on the show, I gotta admit, my feelings were hurt. I thought, God, I'm the Hollywood Jock, how come I'm not up there in cutting-edge black and white? Exactly what does it take to get into this club?

Doh, basic coherency maybe? A proven track record? A couple hundred people beyond family and friends who actually know who you are?

Okay, all right, fair is fair.

We'll just start our own club. We'll get our own show on ESPN. No, make that ESPN2/The Deuce.

"Best Foreign Sports Movies of the Last 6,025 Years."

I can see it now. We'll start with my friend Mitchell Schwartz's entry, *The Ten Commandments*. Final score: Israelites 10, Egyptians 0.

That'll get their attention.

And then, instead of all those professional sports guys, we'll have you guys, the readers. In fact, you can shoot your own video commentary from the comfort of your own homes (or garages if you're married).

(Or doghouses if you're really married.)

Grab a friend, grab a video camera, create your own set and lighting, roll camera, and in the words of Jim Rome, "Have a take, don't suck."

Plus, we'll get some real foreign sportswriters and interview them in their native languages. Only we won't translate it, you know? Just let 'em blabber on in Portuguese or whatever . . . Farsi . . . Mandarin Chinese.

Then we'll take all these brilliant bites of commentary and intercut them with all this great footage from *The Loneliness of the Long Distance Runner* and *Chariots of Fire* and *Karate Kid Goes to Hawaii*.

Just like the big dogs did it.

And you know how ESPN25 had all those sultry model types announcing what number movie was gonna be next? Not just once, but two or three times, very sultry young women with pouty lips whispering, "Number Five." "Number Five."

We're gonna have our own models.

But because we're talking foreign movies, our models will be naked, and they will drive you absolutely crazy as they pout and whisper:

"Der nummer das Suebzegybdvier."

"Ichigou juushishi."

"Numer-r-r-ro venti-quatr-r-r-ro."

Wild, I say. Drive you wild.

Hey, I'm serious about this. I'm calling my agent. "Hey, Guma, get ESPN on the line. No, forget ESPN, get The Deuce. We're bumping billiards right off the band, brother, we're goin' straight for the crotch shot."

I am perfectly serious. Do you realize what a paltry number of viewers a cable station has to grab to be a success?

Out of a country of 290 million people, try 1.5 million viewing households. Hit 2 million and the suits will be dancing; 3 million and it'll be just like the Bushies at Clear Channel kissing Howard Stern's posterior to keep him on the air.

Plus (you're gonna love this) . . . plus we get *Anna Kournikova* to host. (And you know she'll do it.)

You don't think a show like this would generate some heat?

We could do it for nothing, for peanuts. I'll call my friend Steve Kroopnick over at Triage Entertainment (just so ESPN knows some grown-ups are involved). Triage can handle the nuts and bolts, bring on some hotshot MTV editors to give it that slick look, and it's straight to the Emmys, baby.

You can just picture the headlines: "What Were They Thinking?"

Anybody listening out there in Bristol? Do I have to bring this idea to Dave Chappelle? Do I have to, uh . . . uh . . .

Am I foaming at the mouth again? I gotta wrap it up here.

What time is it? I shake my head clear, get my bearings. Hit the "save document" key and pull up my editor's email address.

Should I send it, or should I wait till morning?

Always wait. Last cliché of the day: "Whether it's brilliant or vitriolic, always wait till morning."

But somehow my finger's twitchy on the mouse. Oooops, it's gone. Sorry. My bad. See ya on The Deuce.

34

HOLLYWOOD JOCK FAREWELL

When the call came in from Bristol, I gotta say, I was shocked, shocked! Not that they were pulling the plug, but that they'd waited so long to do it.

See, all year long, I've been flailing away on my column with the very real expectation that any day now, some Big Corporate Kahuna with a thousand responsibilities was going to stumble across my latest effort and say, "What the fuck is this?!"

And the poor guys who actually hired me would start sputtering about real stories from the trenches, insider info, you know, but they'd end up muttering something about how it seemed like a good idea at the time and can they keep their jobs, please . . . ?

It was a weird kind of incentive, knowing I could get the hook at any moment. Go for broke, pal. Let 'er rip. Hey, I jammed in all I could.

And let me make this perfectly clear—ESPN.com has been terrific, right from the beginning and all the way through. I'd like to feel that we've all acted honorably and there truly are no hard feelings.

Like a friend reminded me, "They only gave you your own personal sandbox on the biggest sports Web site in the world for the last eight months."

Not to mention the money.

They even let me tweak them once in a while for crissakes.

Another friend compared the column to a reality TV show without rules. Then added, "Or a slow-motion train wreck."

But it was an honor being on Page 2 with guys like Ralph Wiley (R.I.P.) and Bill Simmons and all the other hotshot writers. Not to mention the original wildman writer of all time, Hunter S. Thompson (R.I.P.).

(Although I guess Homer and James Joyce have a claim to that title.)

Plus, I figured with these guys ranting and raving and cutting it up, I could slip under the radar and just stroll along Ethernet Avenue, shootin' the breeze and selectively flashing open my overcoat—"Hey, buddy, you interested in a screenplay?"

See, what it came down to (and this I understand) was that blurry line between describing the Hollywood process and blatant self-promotion.

With all their other writers, ESPN has rules about this sort of thing. They just weren't being applied to me. Even after being shifted over to Page 3, my stuff somehow kept getting posted.

Now, with the end of the column, I'm really not sure how it's all going to turn out. But there definitely is a book in the works. In the meantime, it was a blast having such an immediate audience; it helped keep me writing.

And I almost made it through the year. This is my 34th column. But now I have to slog out the next seventeen weeks in obscurity in order to complete this "year in the life."

God, there's so much more I want to say to my weekly readership.

About *94 Feet. Zulu Wave. Fourth and Long Gone.*

I'm gonna miss sharing stories about my pal George. How fiery and emotional he can get. How great it is to work with a guy who cares this deeply.

I remember him calling a few weeks ago as we were researching *Zulu Wave.*

"This same phrase keeps popping up on the Internet," he shouted. "*Amandla! Awethu! Amandla! Awethu!* Imagine millions of powerless people shouting it over and over. *Amandla! Awethu! Amandla! Awethu!*"

"What's it mean?" I asked.

"The power is ours," said George solemnly.

"Jeez," I said.

"Either that or, Does that come with fries?"

George knows better than most—if you can't share a cheap laugh, you'll never make it.

And there are still so many more stories from the past.

Like . . . like from 20 years ago.

I remember it like yesterday. I'm hanging around an acting class somewhere down on Melrose, and Tawny Kitaen is in it and she's young and sassy and unbelievably sexy, coming right off of *Bachelor Party* as she is.

(Tom Hanks still can't wipe that grin off his face.)

So Tawny's wearing these tight jeans and a sleeveless man's T-shirt that keeps shifting open as she spins and twirls, acting her heart out on that tiny stage—just so pleased with herself in that endearing, braless manner that'll make a man stupid beyond his years.

After the class is over, I'm hanging around like a dog eyeing a lamb chop, actually following her out onto Melrose as she waits for a ride which clearly isn't gonna show, and I finally gather my courage and offer my services. She looks me up and down, considering, before saying "Sure."

I'm thinking here, Wow, is there a God or what? when suddenly a silver Mercedes coupe pulls up, Marcus Allen at the wheel. Tawny squeals and gives me a peck on the cheek and jumps into the Mercedes and off she goes, leaving me alone on the sidewalk.

Leaving my fragile psyche branded with yet another Hollywood lesson:

Yes, there is a God. His name is Marcus Allen.

This book-writing stuff just won't be the same. One of the great, unexpected perks of the column was making many new friends out of total strangers. And it's been touching, knowing you were out there, loyally groaning your way though my verbiage, week after week.

I learned a lot from you. From all sides of the political spectrum. I learned about respect and dialogue and commonality.

Look, the media will try to fracture us; that's what juices the ratings. But we're all actually much closer than they'll have us believe. We care about family, we care about neighbors. We care about jobs and we care about the air we breathe. We've come a long way in racial relations. (Although our prison system is a total disaster.)

But really, there's a lot to be proud about.

Remember, America has taken on the biggest challenge. You think it's easy to start and sustain a democracy? Hey, let's really make it tough and throw in just about every disparate national, racial, economic, sexual, religious, and cultural group you can find under God's sun. But it's working.

And we really have come far when it comes to tolerance.

The other night, I watched *Queer Eye for the Straight Guy* with my 12-year-old son. We laughed. We howled. It was no big deal. Can you imagine doing that with your parents when you were a kid?

That's progress.

I also learned how we all share a deep concern for our troops overseas. I'll say it again: Say a prayer for these guys, and let's figure out the best way to get them home.

And for my new friends from other countries, other continents, you remind us that we're all in this together. On a small, crowded planet.

All right, I'm almost out.

But get this, ESPN.com has offered me a freelance gig. Anytime I want, I can pitch them stories for Page 2 or Page 3.

So if you have any ideas, pass 'em along (but let's not get too wacky here, I'm still hanging on by my fingernails).

I've asked Bristol to make my email address hot—robryder@adelphia.net.

Actually, how about *everyone* just popping me an email? Then I'll have a way of getting in touch about the book and other stuff. (Will this crash my computer?)

I might even start my own blog, who knows? It'll be www.hollywoodjock.com if and when the time comes (but blogs are dangerous animals, right, Gregg Easterbrook?).

I end my last column with this thought:

We turn to sports for escape and for comfort. And out of respect for the game.

But if we really respect the game, we also know when to turn it off. We know when to get back to the task at hand—saving the world for our kids. For all the kids.

That way, when we turn back to the game, it'll still be there.

We'll see ya around. Ciao. Adieu. Farval. وَدَاع. Auf Wiedersehen. Прощание. I'm outta here.

Amandla. Awethu.

35

GUYS DON'T BUY BOOKS

So my last column hits the Internet and immediately the emails start pouring in.

"Dude, you got waxed!"

Hey, thanks a lot. But actually, the words are strangely comforting. Jock-speak. He's kind enough to add—"Tremendously disappointing news."

"Rob, you are the thousandth point of light," reads the next email. "And fuck Marcus Allen." (Thanks, Todd.)

And then—"*Zulu Wave* sounds like it has some legs. I worked in movies as an inventory analyst for Best Buy for about three years. The movies you are pitching, if they can get produced, will sell on DVD. Trust me they will sell. As long as a turd is priced right and made to look pretty, we could sell it." (God, there they go again. What kind of readership did I have, anyway? Thanks a lot, Brad.)

Okay, there are hundreds more (and believe me, it's tempting to simply cut-and-paste my way through this writing life) but, here's the last one:

"Mr. Ryder, I have enjoyed all your columns. I too like to write screenplays and skits, but just for fun (plus they suck too). Best of luck on your future endeavors." (Thanks, Arthur, and just remember, if *Gigli* could get produced, so can you.)

So now what?

I've got to convince myself that losing the column is the best thing that could have happened to me. It was wearing me down. Especially the fights with my editor. I was spending four to five hours every Tuesday once the column was posted just trying to get the typos fixed. It was driving me nuts. And I was running out of gas.

So, yeah, forget the fact that I was actually pulling in 2K a month from the comfort of my own living room. And forget the fact that in a life filled with uncertainty, vagueness, and dubious self-image, the one gig that had actually given me an ongoing identity was now gone. I gotta suck it up. I gotta face facts.

Today is the first day of the rest of my life.

I spend it floating around on a raft in the pool, drinking too much beer. And the next day. And the next. And yeah, we're renting a house with a pool, okay? You know, I might drive an 18-year-old Toyota with 187,000 miles on it, but we're living in a house with a pool. For all you guys out there who are slaves to your car payments, I have just one word—reconsider. Me, I'd rather park around the block than swim at the Y.

But clearly, I am fucked. Living on borrowed time. Then, miraculously, this email from Guma pops up:

"Talked to McMahon [the editor at William Morrow]. He's putting together his offer, but just so you know, he sees this as a trade paperback original—thinking, as he and sales do, that for this book to perform well it needs to be priced attractively. They don't think this book will find much of an audience priced at $25." (Hey, who do you think's been reading my column, a bunch of cheap bastards who'd rather ride around on leather seats than read books?) Guma's email continues: "I tend to agree, as you're going after guys who don't buy books in the numbers they should." (See the contempt these literary geniuses hold you in?) "That said, this means you'll most likely get an advance in the neighborhood of $25,000. You cool with that?"

Am I cool with that? A 25K advance? For a book that's two-thirds finished? Yeah, I'm cool with that. I burst upstairs and share the news with my wife. She's happy for me. I guess. Because I'm convinced I see that "Can't you just call it quits and get a real job?" look in her eyes.

"Really," she says. "I'm happy for you. When do you start?" (Which really means when does the check arrive?)

"Uh, you know. If it's anything like Hollywood, it's gonna be weeks."

"So maybe in the meantime, you can get your life in order."

I stare at her dumbly. We've been married for 18 years. (In fact, 18 years ago, as my Toyota Camry was rolling off the assembly line in Kyoto, Japan, my wife and I were exchanging vows on the upper deck of the faux paddle wheeler the *Cherry Blossom* on a stormy day on the upper reaches of the Potomac River.) And here's my point. She's known me almost 20 years. Yet somehow she's earnestly suggesting that I can discard my past and reorder my entire life—years and years of deep-rooted neuroses, barnacle-encrusted habits, time-tested, carbon-dated, foolproof methods of procrastination—rework the whole shebang in a matter of weeks.

"Oh, honey," I say.

"You can start in your mother's garage," she answers blithely.

Oh, no. Not my mother's garage. Somebody grab me the garlic and crucifix. Look, it's not my mother that puts the fear of God in me. It's my mother's garage. My mother I can deal with. Really. As long as we don't talk about God, religion, sex, abortion, or property taxes, my mother and I get along just fine. It's just that every chance she gets, she tries steering the conversation back to one of these topics. I mean, we can be having a perfectly amicable chat about how anyone who makes less than $150,000 a year and votes for George Bush is a total moron, when suddenly she's saying "But you know, Rob, there are many facets to social justice."

And I say to myself, Oh, shit, here it comes—and suddenly I'm up out of my chair. "Mom," I shout, "I think I left the gas on in the garage!"

This throws her long enough for me to make my escape.

It's a funny thing, but I've actually grown to appreciate living so close to my mother. I get to cut her grass on Saturdays. Bring the boys over to pull weeds and rake leaves. And as my wife continually reminds me, it's a chance for me to model behavior of how boys should treat their mothers when they get old. My wife is no dummy. And seeing how she, like many mothers, has been cursed with only boys, she's interested in seeing them receive at least a rudimentary education in family responsibility. Face it, in our society, it's the women who get saddled with taking care of elderly parents. Except when there aren't any. My brothers are living in Bozeman, Montana. Thanks a lot, bros. So it's on me. Sort of. But like I said, it's not really a burden.

Besides, it's not my mother, it's all my shit in my mother's garage. Boxes and boxes, reams and reams of shit. Scripts, treatments, contracts, videos, business plans, old photos, old love letters. Boxes rotting from water damage. Boxes thick with spiderwebs and mice droppings.

I swing open the garage door. Step inside and turn on the light and reach for the first box when I'm suddenly paralyzed. It's too much. I can't do it. All the scripts that never got produced. All the treatments, the notes, the once brilliant ideas that never saw the light of day. It's overwhelming. If I'm not careful, my mother's gonna come out and find me crying over a box of mouse turds. I can't do that to her. I can't do this to myself. So I do what any self-respecting, arrested-development, namby-pamby baby boomer would do in a case like this: I escape back out into the bright sunlight and the refuge of the raft in our pool, the comfort of yet another beer in my hand.

36

"I'LL BELIEVE IT WHEN I SEE IT"

I'm getting waterlogged. I gotta get off my ass. I shuffle through my projects like a three-card monte dealer working the rubes from New Jersey. (Wait a minute—I'm from New Jersey.)

Holly & Vine. Remember the series my pal J.B. White and I pitched at Lifetime? Whitebread actress and hip-hop P.I. team up to solve crimes in Hollywood's underbelly? This is a good idea. This idea isn't going away. We've got Aaron Lipstadt, a legitimate television showrunner, hooked into the project, and we want to keep hustling it.

I call J.B. His agents at William Morris have shown our treatment to Morris Chestnut's agents and it seems there's some interest there. All right. William Morris, Morris Chestnut. Morris Chestnut is a talented actor. And it seems that these agents are packaging. They're really, really packaging! But pitching season is almost over, so we've got to move fast. No, correct that. The agents have to move fast. The writers have to wait, because that's what writers do.

I ask J.B. how his other stuff is going. J.B.'s been making a credible living as a television movie writer, but that world has been shrinking with the explosion of reality TV. Nonetheless, he's got a movie set up at USA that just might get made—*The Twelve Days of Christmas Eve.* (Think *Groundhog Day*.) J.B.'s update—"They're still trying to cast the movie.

Michael J. Fox—too sick; Jeff Daniels—too expensive; Alec Baldwin—too not interested; Chris O'Donnell—too young; Dan Ackroyd—too not what the network wants; waiting for the next name. Production office has been set up in Edmonton, where it will be shot; locations are being scouted; production designer hired. Sound like we're making a movie? I'll believe it when I see it."

Now that is the attitude of a survivor. Of a winner. That's the attitude I'm trying to reassume.

When you live a freelance life, there's one rule of thumb—get it out of your ass.

Okay, it's out, it's out. What's next? *ZuluWave*. I email David Madden at Fox Television. He loved the pitch. He said he'd talk to his old partner, Bob Cort, at Paramount about it. Remember, Cort has a discretionary (slush) development fund. If he likes something, he can write a check without studio approval.

But I haven't heard from Madden in six weeks. Six weeks of unanswered phone calls and emails. And Madden's one of the good guys!

Meanwhile, the young hotshot English producers are about to make an offer. I'm reluctant. It'll be real lowball. Hardly worth it. And suddenly we'll have to answer to them about how the story goes.

Remember, as soon as you accept money from somebody (even a piddling amount), they own you. They own your project. You answer to them. I say we wait for Madden. He'll call. I know he'll call.

He doesn't call.

Then a rather terse email arrives from George in which he recounts that David Madden has gone "cold as a corpse" and that my other connections "seemed to disappear after the first blush of enthusiasm," so he wants us to seriously consider the offer being made by his English producers even though it may not be all we want.

Okay, George, thanks for sharing. I call him. "Let's see what they have to offer. I'm not gonna hold out for some pie-in-the-sky Hollywood development deal here. If they're willing to pay a bit up front and seem committed to making the movie, let's go with them."

See, it really is the economy, stupid. My wife's last paycheck dried up weeks ago. I need some quick cash so I call Michael Knisley, my first editor at ESPN.com. "How about a story about the Southern California Summer Pro League? My pal, Clay Moser, is running it this year and there's gotta be some real human interest down there."

Knisley gives the okay, so I head down to Long Beach State, down to one of the really cool gyms—the sleek, shiny blue Pyramid, seats about 4,500. Clay's left a press pass for me (that's a first) and I wander in. The place is nearly empty. I sit up high and watch a bunch of desperate, wannabe NBAers trying to impress the NBA personnel scattered around the floor.

Is there a story here? There are a thousand stories, and they've all been told before. Jesus, what have I gotten myself into? How am I gonna fashion something fresh and different out of the same old same old?

I spot my old coach at Princeton, Pete Carril, and get that sickening thud in my stomach. But hey, Pete makes for good copy, and I promise myself I'll get around to him before the day is through. Then I spot another old warhorse, Hubie Brown, who's also a blast from my past. I know Hubie from when he was coaching at Fair Lawn High School (New Jersey) and I was playing for Paramus. Hubie went on to become one of the craziest coaches of all time. Ask anyone. He would scream, foam at the mouth, rant and rave like the total absolute lunatic he was. But, God, he knew his basketball. And for some strange reason, you couldn't help but like that guy. He put it out there. As long as you didn't have to play for him.

By my senior year in high school, Hubie was coaching down at William & Mary, and he invited me down on a recruiting trip. The Athletic Department arranged for a couple other recruits and me to be escorted around by some hospitality dates. When these three poor freshmen coeds showed up, Hubie winced. They weren't exactly beauties. Hubie pulled me aside and whispered, "Jesus, sorry, what a fucking disaster."

But hey, at least we were going to a concert that night. With any luck, it'd be the Blues Project, or Chicago or B.B. King or, I don't know, anything but Up with People. That's right—Up with People. I about lost it. The concert started with three long-haired hippies sitting around onstage playing acoustic guitars and singing protest songs, and that fast, a huge group of squeaky-clean white Amerikan Youth came marching onstage singing "God Bless America" and literally ripped the guitars out of the hippies' hands and shoved them off the stage. This was 1967.

Anyway, I spend the day wandering around the Pyramid, talking to Hubie, Pete, a player from Jamaica, a recruiter from Korea—"We got Outback, we got Wendy's, we got McDonald's." Plus a bunch of other people and then I grind my way home up the 405 and plunk myself down at the computer to see if I can work some journalistic magic.

I can't. I just can't do it. I'm not motivated.

Then this email pops up from Aaron Lipstadt, the showrunner from *Holly & Vine*.

> JB, Rob,
>
> I talked to Bill [his agent] last night. He said he's having a hard time setting up pitches for H&V—he's being told it's too close to *Veronica Mars*, a new series on UPN I think, about a high school grad girl who goes to work in her dad's PI agency! Apparently the "vibrant youth joins forces with the older law enforcement pro" card has been played, race issue or no, and buyers are reluctant to feel like copy cats. I'll keep on him, I know he's still working it. Any ideas?
>
> Aaron

Yeah, here's an idea. Why don't I get to work on the Summer Pro League article because shit like this keeps happening to me, and I'll probably never work in Hollywood again? So I knock out an 800 word slice of life about the Summer Pro League, and Knisley likes it and two days later it's up on Page 2 on ESPN.com and just like that, I'm a journalist with a fresh 500 bucks in my pocket.

37

IF THE TRUTH HURTS, LIE

What do I do if I don't score? When you're 54 years old and you need to reinvent yourself, the choices are limited. You can just hear the gal in Human Resources—"And what have you been doing these last 30 years?"

The alternative will be starting a business. Anyone can do it.

Clay Moser calls. The owners of the Summer Pro League are quietly looking to sell it. "It could be a little gold mine," Clay says. "There are so many opportunities to make money outta this thing. . . ."

I'm intrigued. For one, Clay is about the most solid guy I know. He has years of experience in management in the Continental Basketball Association. And he's been in great helping develop our 4Man Basketball League business plan. And as I learned the hard way, start-ups that require a lot of capital are much tougher prospects than the purchase of existing businesses.

The SPL could be perfect. It's got some of the biggest brand names in sports (Lakers, Knicks, Mavericks) ACTUALLY PAYING to play during its two-week summer schedule. It's in a cool venue. It's well-established. It'd be a great laboratory to try out some of my ideas about hoops and entertainment. Plus we can use it as a springboard to launch the 4Man League.

I decide to drop everything and make a run at this. Fuck Hollywood; it's all too iffy. Here's something I can sink my teeth into.

The next few days are a whirlwind. A series of phone calls, then several meetings with the owners where I assemble as much financial information as I can. The trick here is to come up with a valuation of the company that's realistic. And one that's based on performance, not potential. Then I spend a couple days locked into a spreadsheet on my computer. On one side there's revenue—tryout camp, team entry fees, ticket sales, sponsorship sales, etc. On the other side, expenses—venue rent, uniforms, advertising, salaries, travel, etc. A hundred categories that must be logged and configured.

I'm digging this. Hard numbers, spreadsheets, math. No worries about Act II going squishy (as they inevitably do). And the numbers add up. This thing has made money. And is capable of making considerably more. Six-figure profits. I call Clay and we talk it through. He's got a couple of investors who've expressed interest. I've got a few guys myself—including Ice Cube's partner, Matt Alvarez. I call Pookey Wigington and talk it through with him. He's intrigued as well—being a former baller at Seton Hall and current producer. So I feel secure in opening a negotiation.

I figure the SPL is worth in the middle to high six figures. It would be more, but there's a huge downside to the business—none of the NBA summer teams that play are guaranteed to return year after year. Plus there's a new summer league that just launched in Las Vegas, and they're aggressively poaching teams from the SPL. No way the current owners are going to be asking too much for the thing. They'll never get it.

So I grind my way down the 405 to their offices on Olympic in West L.A. We exchange hellos, sit at a conference table, and I look them in the eye and ask, "So what do you want for it? Where do we start?"

"We think it's worth up to 10 million dollars," comes the answer.

Oh, man. I feel that sick soft thud in the pit of my gut. Another dream gone bust. Boy, that was fast. I stagger out into the blinding sun on Olympic Boulevard, climb into my too-many-miles Camry, and lurch my way home.

So once again, it's back to the drawing board. Back to the home office to pick up the pieces. But I'm feeling strangely rejuvenated. I made a credible run at a business opportunity. And who knows, maybe it's not over yet. I give myself a pep talk. I somehow manage to be my own Tony Robbins. I've got a lot of great things happening. The book, some very real movie possibilities, a future as a sports journalist. Who knows, maybe (as one reader suggested) I can make a lot of dough on the motivational speaking circuit. My motto, the basic core of my beliefs: Rule #1— If you don't feel it, fake it.

But when I walk through the front door, I'm confronted by my wife.

"How'd it go?" she asks.

"Fine," I lie.

Rule #2—When the truth hurts, lie.

I slip downstairs into my basement office. An email's arrived from Will Staeger, who works for ESPN Original Entertainment. "Congratulations on Jock! That's awesome. . . . Man, I'll tell you, Matthew can sell his ass off, can't he?" W

See, everything's gonna be all right. Besides working for ESPN television, Staeger is also a writer—he's got a couple of novels under his belt—and Matthew Guma's also his agent. It's a great connection. I've pitched him *94 Feet* over the phone and he's interested. But I told him I wouldn't send him the script until I've got some elements attached—a producer or director. A name actor. A group of young NBA stars who are willing to play college guys. (But to use a tired expression—that's like herding cats.) Nonetheless, Staeger's email is a positive sign.

I hang up and email the producers Mark Burg and Oren Koules at Evolution Entertainment. Burg I've got my doubts about (no matter how you slice it, *94 Feet* will never be a financial windfall). But his partner, Oren Koules, is intriguing. For one, he's a former hockey player. Plus he owns a minor league team up in Montana. He just might get it. That quick an email flies back, "Send the script."

Sweet. And the feeling of momentum is buoyed when suddenly an email from Pepper Rodgers pops up suggesting that I might help him with *Pepper 2*, the second installment of his autobiography. Okay, so that's something. That's legitimate work. But do I want to spend six months of my life listening to (then regurgitating) wonderfully funny stories about the real-life antics of a colorful old college football coach? I don't think so. Books are a bear. I once helped out a friend of mine, Byron Laursen, who was helping Pat Riley write his second book, *The Winner Within*. My part was basically a cut-and-paste job. I was spending days down at the Hollywood Public Library, pulling quotes and articles from old newspapers and magazines. Then Byron and Riley were going to weave them into their narrative and that'd be that. But you know what? It was a lot of work. Writing a book is digging a trench. Even a shallow trench like *The Winner Within* is still a grind.

Nah, I've already got one book to write. I wanna make a movie score. I remember *Zulu Wave* and the hotshot English producers who are about to make an offer. Yeah, it'll be modest, but still, a little cash in hand . . .

I call George. "You hear from them?"

"Yeah," George answers in an uncharacteristically muted voice.

"They make an offer?" I ask.

"Yeah, they made an offer," he answers.

"What'd they offer?" I ask.

"They offered to be the first ones to read it once we've written it on spec," says George. I say nothing. Then George lets loose an extremely creative burst of invective and I can't help but join him—there's screaming back and forth about the scatological fate these hotshot English producers truly deserve—invective that I can't even quote here because after all, George and I are well-respected members of the community and everybody knows rage has no place at the table.

Rule #3—When pissed off, rant and rave like a lunatic; it makes you feel better.

George and I hang up, pledging that we're gonna spec *Zulu Wave*. Suck it up and write the thing. This tale deserves to be told. That night at the dinner table, I can't bring myself to tell Andrea, instead placating her with promising stories about the realities of sports journalism. It can be a steady paycheck, for God's sake. And you only have to subject your manhood to whiny, spoiled millionaires three, maybe four times a week, which is better than your average working sap.

The dinner's delicious. Our two boys are lively but well-behaved. Life could be worse. Late at night I'm taking out the garbage cans to the street. A fog's rolled in off the ocean. The trees and parked cars hulk in the shadows. I hear voices. I peer into the thick mist. Slowly a group of teenage boys emerges from the fog. It's freaky, mesmerizing.

"Hey," I call out.

"Hey," a kid answers.

The group starts to slide by. I think I recognize a couple of them (I know most of the kids on the block) but I'm not quite sure. "You look like ghosts," I say.

One of them answers, "Yeah, and you look like a fag."

They snigger. Oh, fuck. Do I have to deal with this right now? Why do I have to deal with this right now? "What'd you say?" I call out.

"You heard me," comes the answer.

I suddenly find myself striding toward the group. "Yeah, well, let's hear it again! Let's hear you say it to my face!"

The group begins to stride off up the street. I quicken my pace, they start to jog.

"What are you runnin' from?!" I shout.

"You, faggot," the kid shouts. I break into a sprint. They break in all directions, dashing between parked cars, sprinting up driveways.

I stop. Watch them melt into the night. "Chickenshits!" I yell.

"Faggot!"

"Chickenshits! Go home to your mothers!"

38

A MIDSUMMER NIGHT'S MESS

I write another article for ESPN.com. Magic Johnson's charity game down in the Forum. It's actually two games. First the Hollywood celebrity game—Boris Cojo, Duane Martin (from *White Men*), Larenz Tate, Kevin Hart, and, of course, "The Game," who's a hot young gangsta rapper outta Compton, part of G Unit. These celebrity games used to be fun affairs—lots of clowning around. But that's changed, probably because of the NBA Entertainers League, where these guys have been working up some major rivalries.

So the celebrity game is hard-nosed, fast and furious. The equivalent of high-level high school (which is saying a lot—hoops is a tough game). But what still gets me—and you can see it in their eyes, in their demeanors—these guys really think they've got the goods. What they don't understand, if you up the competition just one notch—say to junior college or Division III—there are a thousand guys who can shut you down. They'll take away your dribble. They'll block your jump shot. They'll bump you off the ball, they'll jam you up, they'll box you out, and then they'll dominate you on the other end. You come away from an experience like that with your head spinning. Schooled.

But hey, it's cool. It's for charity.

Surprisingly, the second game, the NBA guys, goes pretty much the same. See, as the celebrity games got fiercer, the NBA games had become more fun—lots of showing off. Lots of space out there. But for some reason, today's game is different. Today it's the veterans—Paul Pierce, Antoine Walker, Gilbert Arenas, etc.—going up against the rookies, represented by T.J. Cummings, Chris Duhon, J.R. Smith, and the like. The rookies decide to play hard. The veterans respond in kind, and that quick we've got a game on our hands. All right, this is how it oughta be.

Then, suddenly, The Game comes striding out of the locker room to watch. White T-shirt, baggy jeans, jewelry hanging every which way. Tattoos on his neck. Now that's pretty hard-core. I could imagine someone sticking a tat needle in my arm, maybe my back even, but my neck? No way. You gotta really need some pain to ask for that.

Anyway, The Game somehow ends up sitting next to me courtside, surrounded by a gaggle of wide-eyed kids seeking autographs. I introduce myself, throwing the ESPN name out there to serve as the automatic legitimizer that it is. It works. We shake hands. I consider asking him whether he prefers to be called "The Game" or just "Game," but then I remember that this guy's been shot like 19,000 times already and after that went out and got his neck tattooed, so instead I comment on his watch—which could easily send both my boys through college and then some—and he holds it up for me to see. Wow, lots of diamonds—lots of South African miners owing their livelihoods to this hip-hop gangsta in America. But when I offer to trade my Bic pen for the watch, The Game suddenly goes cool.

I bomb home late after the game, and this time the article flows pretty easily. Unfortunately, it's bound for Page 3, not Page 2, and suddenly I'm back in the throes of my old editor, who in a burst of frenetic activity decides to reorganize my piece, somehow combining the two games into one, expunging my funniest stuff and contributing a multitude of typos in the process. And then this editor doesn't even have the respect to run it by me for posting it. The next morning, I go to the Web site. It's up. Oh, fuck.

I fire off an email to the editor's boss.

Rule #4—Never ever go above your immediate boss's head unless you don't give a shit.

My editor's boss is apologetic. He intimates that this editor will soon be getting nudged to the sidelines. He asks that I bear my way through this one. I then get an email from the editor, who is begrudgingly polite, and we get the thing fixed. (And believe me, the article is no gem, but now at least it's fairly intelligible.) But I spin away from this latest foray thinking, I don't think I'm cut out for this journalism shit. As a screenwriter or book writer, you've at least got these long breaks between getting dumped on by people.

Clay Moser calls. Disappointed that the Summer Pro League deal is going nowhere. "Maybe they'll come around," says Clay.

"Maybe they will," I reply. In the meantime, Clay's got an interesting proposition. He got an offer from a promoter to assemble and coach a team of American ballers on a tour of China next month. I think, There's a nifty documentary. Clay gives me the number of the promoter. We hang up and I sit back for a minute. Is this something I want to get into? It's such a weird time right now. What do I want the most? I want to be a highly paid screenwriter. Oh, that's all? All right, barring that, what's next? Documentary filmmaker? That's pretty cool. I decide to make a run at it. But the smart way. There are plenty of guys out there making quality docs. They've got the experience, the relationships, they know the marketplace. I decide to pursue the project only if I can attract a top-notch production company.

So I call the promoter—a Chinese-American immigrant who speaks broken but intelligent English—and we hammer out an understanding. I talk to Clay again. How reliable is this guy? What are we getting ourselves into? Clay gives me a list of players who toured with him the previous year.

Aha, Cedric Ceballos. I worked with Ced on *Eddie*, plus he's great friends with Pookey Wigington. They played juco together at Ventura

College before Pookey went on to Seton Hall and Cedric to the NBA. I call Pookey, explain the situation, he gives me Ced's number, and five minutes later:

"So Ced, tell me what it was like."

"Total disaster. It was a mess, man. The hotels, the travel, guys getting paid, everything. I was paying for my own hotel, man. And the food—wow."

I thank him for his time, hang up, and think, Okay, now we're onto something. Reality TV, baby. Conflict is everything. But no way I'm moving forward without a production company to do the heavy lift ing. I call Brett Rapkin at Black Canyon Productions.

"Have you got something written down I can show my boss?" he asks.

"You'll have it in an hour," I reply.

I hang up and bang out a quick three-page treatment. *China Slam*.

Brett says he likes it. Wonders if the tour will attract any celebrity names. God, here we go again. I explain that the whole point is how these guys are all nobodies. The second- and third-tier guys. The guys who've gotten lost in the shuffle. Brett says he'll run it by his boss.

Okay, that's cool. In 24 hours I just put another horse into the race. And the beauty of it—there's a two-week turnaround on this one; it'll be a go or a no go, and I won't be sweating it, 'cause it was a snap to put it into play.

I fall asleep thinking, Man, I might be goin' to China.

39

FROM $25,000 TO $6,375

Looks like I'm not going to China. Clay found out that all along this promoter's been hustling another coach for the same job. And one thing you learn fast about Clay Moser: You mess with him, he walks. Plus he's not looking to make some fakakta Murphy's Law documentary—he's looking to coach a basketball team and get paid for it. At the same time the honcho at Black Canyon passed. All right, so it goes. I've got other fish to fry.

I get a phone call from Oren Koules at Evolution Entertainment. He wants me to come in and discuss *94 Feet*. He connects me with his assistant, Chad, and we set the meeting for next Tuesday, 11 A.M. I hang up thinking, Aha, so what's this all about? This is feeling like something real. Oren Koules seems like a no-bullshit kind of guy. Still, I'll be surprised if the meeting isn't postponed. All meetings are postponed. But hey, Koules says he wants to meet. This is cool.

Plus, I've got other projects to push forward.

Like *Hollywood Jock*. I actually got the contract from my new editor at William Morrow (a division of HarperCollins). Only now, the $25,000 advance we asked for has become 20 grand, which still ain't too shabby, but I only see half up front. And then, out of that 10K, Guma and Inkwell Management gets $1,500, plus out of the remaining $8,500 I've got to put away at least 25 percent for taxes, and suddenly that $25,000 advance is looking more like $6,375. Still, it's real. And it's a deal. And most important, Hollywood studios buy books to turn into movies.

I spend a few days grappling with the problem of how to turn this book into a movie. My solution? Throw the book away. Or most of it at least. Create a black-white buddy comedy. Two ballers who meet at the Hollywood Y. Both looking to write, act, and direct. Both ending up working as sports advisers to the movies. Working on the same shows sometimes, just like me and my real life friend Kevin "K to the B" Benton. It feels like a natural. Everyone's fascinated by the movie business. Combine it with sports, work that black-white buddy thing, and it's bling-ching-aling all the way to the bank, baby.

I work up a pitch and call Guma, who helps set up another meeting with Matt Alvarez at Ice Cube's Cube-Vision at Revolution Studios. Unbelievably, the meeting's for tomorrow, and it doesn't get postponed.

Revolution Studios resides in a sleek, sterile building in West L.A. on Olympic. I park and sign in at the lobby desk and head upstairs to a conference room, where Matt Alvarez soon joins me. Alvarez is young and smart, with a laid-back manner and wry smile. He's got a beard and short hair and he isn't playing the Armani game. He's dressed down and hip.

This is the new Hollywood.

A couple months ago, Alvarez's name came up in a meeting with an entertainment lawyer, and this suit (he'll remain nameless) said, "Yeah, Matt's just like the guy you bought dope from in college."

And I sat there thinking, What kind of thing is that to say about somebody?

Besides, the resident dope dealer when I was in college wasn't anything like Matt Alvarez. He was a former all-state wrestler, the kind of guy who could knock out 18 perfect chin-ups without being winded. He called himself Wink and wore blue-jean overalls and wire-rim glasses and had a huge frizzy Isro (the Jewish counterpart to the Afro).

Isros were a big thing back then. About half the 1969 Harvard basketball team had them. I remember going up to play them on a freezing Friday night and there were about 11 people in the stands and when the Harvard team came out, they had Isros out to here and they were all laughing through the layup line, and our crusty coach, Pete Carril, muttered, "Jesus Christ, what a fuckin' world."

Of course we killed them, but Carril was still spitting invectives about our sloppy play. I remember walking off the court and slipping into some philosophical quandary about why we were actually putting ourselves through this grief: So, you just slaughtered a couple of phenomenologists and a Kierkegaardian, how's it feel?

Anyway, the entertainment lawyer's crack about Alvarez stunk of jealousy. Alvarez is a young guy who's partnered up with one of the hippest, hottest actor-producers in Hollywood. Cube can open any door. People want to be in business with him. Part of Alvarez's job is making sure they pick and choose wisely.

"Look," he says, "we've got a few feature projects set up, an animated TV thing, a show based on the history of hip-hop. And we get approached all the time—say a "life in the hood" reality show, something like *Pimp My Ride*. But that doesn't make sense for us. If you think of Ice Cube as a brand, spreading him around like that you're just diluting the brand."

Ice Cube. Ice Cube the Brand. Twenty years ago, he and Dr. Dre and Eazy-E called themselves NWA (that's Niggas with Attitude, fyi) and they released a song called "Fuck Tha Police." Now here he is, trying to protect his brand. Just like in the words of my former coach, "What a fucking world."

I'm gonna get killed for saying this (screw it, I'm not running for office) but "Fuck Tha Police" was a legitimate call for justice. The cops in south-central L.A. were systematically harassing and humiliating young black males, and if you don't view that song as a cry of outrage, you probably think Dick Cheney is a prince.

In fact, through researching *Zulu Wave*, we learned that in the 1980s, rebellious young blacks in South Africa listened to a lot of hard American rap. Did Ice Cube, in 1988, realize his music was fomenting one of the greatest racial revolutions in world history?

I'll have to ask him.

Matt Alvarez is joined by John Hayes, a calm, insightful African-American. We talk some more—about Hollywood, the familiar litany of how hard it is to get things made.

My view—you've got too many executives with too many choices. The whole town gets paralyzed.

Alvarez says, "Michael Eisner's kid was in here. He and a buddy had a movie idea. So we said, Yeah, write the script. And they did and brought it back and it wasn't half bad. But then it was Yeah, all right, now rewrite it. So they did. And they came back and they were like, Okay, let's make this movie now. Like it was that simple."

We all laugh. As much at ourselves as young Eisner. It just ain't that simple.

Then we talk about Cube's latest movie, *Are We There Yet?* Alvarez and Hayes speak positively about the January release of the romantic comedy. But me, I'm skeptical. I've seen the trailer—it looks soft. Of course, I don't say as much. But I seriously doubt Ice Cube's gonna generate the crossover appeal they'll need to even break even. If this movie makes 20 million domestic, I'll be seriously surprised.

Better get off the subject before I say something stupid. Once a film is locked and before it's released, there's a grace period where you don't want to jinx anything. The critics and public will pass judgment soon enough.

Finally, the conversation turns around to *Hollywood Jock*. So I go into a six-minute pitch.

When I finish, Matt Alvarez and John Hayes stare blankly. I knew I lost them halfway through—no, that's not right: I knew I never had them. There's no sense in beating around the bush here. If you pitch something and it doesn't fly, find out why. So I ask.

"It's a tough sell," Matt Alvarez says truthfully. "Movies about movies. That's tough."

I think about bringing up *Get Shorty* or *The Player*, but decide against it. When an executive is honest enough to say why he's not interested in something, go easy on him. Or else he'll turn into the typical guy who'll give you that "God, that's great" line right out the door and you'll never hear from him again. John Hayes speaks up to say there's a movie about moviemaking being released in a matter of days, called *The Last Shot*. He says let's see how that does.

We shake hands good-bye and I hit the freeway. Fighting this latest disappointment. After 30 years I finally get a book deal and what's it about—the one subject movie studios aren't buying. Nicely done. But hey, what if *The Last Shot* opens big? The whole thing will turn on a dime.

I get home and Google through the upcoming releases. *The Last Shot*: A movie director (played by Matthew Broderick) finds a man to finance his latest project but soon discovers that the producer (Alec Baldwin) is actually an undercover FBI agent working on a mob sting operation.

Hey, that's not a bad premise. It's actually based on a true story. It opens this weekend, and I head into Friday with that clutch in my stomach like *The Last Shot* is my movie—and it might as well be, with all I've got riding on it. But I just know it's gonna be a hit. I know it.

40

I SURVIVED BOBBY KNIGHT

I wake up Monday to learn that *The Last Shot* bombed. The marketing guys at Universal must have known it going in from their focus groups because they only opened it in six theaters. Cut their losses. They'll let it limp along for a couple of weeks, then they'll pull it and it'll slip into movie oblivion.

Next.

I've got a meeting tomorrow with Oren Koules of Evolution Entertainment about *94 Feet*, and it's as important as any meeting I'll have all year.

So today is not a day to work on *Zulu Wave*. Today, I'm gonna leave *Hoop de Ville* on the shelf, leave *Hollywood Jock* buried in my hard drive, and get to work on *94 Feet*.

I want to go in to Evolution prepared. I reread the script. It's there. I can see this movie. God, it's great to have a script you believe in. I've put in the hard work—now's the time to make it pay off.

I pull up a file—a possible cast list I've created. I'd love to see Lou Gossett Jr. as the wily old black coach. Morgan Freeman is on the list too, but that's like asking God to be in your movie. Then there's Ossie Davis (who has since died. R.I.P. to one of the great black men of the last century, a man who led by example). Or how about a stretch, Garrett

Morris of the old *Saturday Night Live* days? There's no actual model for this character. John Chaney, basketball coach at Temple, comes closest, but unlike my character, Chaney loses his cool. In a press conference once he actually screamed at then UMass coach, John Calipari, that he was gonna kill him. Now this isn't quite as bad as Mike Tyson saying he's gonna eat your babies, but still . . .

No, my character's the king of cool, calm, and collected. I've known older black men like this, and they are very impressive. Imagine what they've been through. The '50s. The revolutionary '60s. Imagine their strength and perseverance as they navigated the roiling waters of racism, year after year, decade after decade. Maintaining their dignity and pride. Feeding their families. Achieving success against the odds.

So, Lou Gossett Jr. to bring this character to life.

And how about Ed Harris as his fiery white counterpart. Fiery? Did I say "fiery"? This character is based in part on Bobby Knight. How about lunatic? How about screaming meemie? How about petulant whiner? How about brilliant bully? (Knight's been offered his own sitcom, btw.)

Bobby Knight is not my kind of guy. I don't care if he makes sure his players graduate. I don't care if he's won three national championships and understands the game of basketball as only a handful of others on the planet.

So why do I despise Bob Knight so much? Because the man treats people like shit. He's created a culture of fear that surrounds him wherever he goes. He walks into the gym, he walks into the chancellor's office, he walks down the freakin' supermarket aisle and people flinch and cower. "Oh, God, what if he goes off? Please, don't let him go off on me. Please . . ." Eyes go to the floor, twitches erupt, articulate men stammer and mumble.

Look, here's what I don't get. And here's the rub with boys growing up to be men. When you subject yourself to a coach like this, you are violating the first rule of being a man—YOU DON'T TAKE SHIT FROM ANYBODY.

And I'm not talking about some coach getting right up in your grill for missing a blocking assignment. And I'm not talking about some history teacher giving you a C for a mediocre effort. Or a boss making you do it until you get it right.

I'm talking about allowing yourself to be demeaned and humiliated. What's the life lesson in that? That's not great preparation for the real world. Hey, great, I just survived Bob Knight, now I'm ready to go out and spend the rest of my life being trashed by every boss, wife and Joe Schmoe out there.

Or worse, I survived Bobby Knight, so now I'm free to pick on every subordinate, every gentle female, and every snot-nosed kid in my path. In the real world, believe it or not, people are mostly civil to each other. The best bosses are understanding. The best workers are attentive and willing. The best spouses know the give-and-take of a lasting marriage.

Sure, there are screamers. They're a pain in the ass, but a major life lesson is learning to duck and weave when you see one heading your way. It's jujitsu, baby, using their power to propel them on their twisted journeys.

You think when James Earl Jones was a young man he put up with Bob Knight's kind of crap? How about Martin Luther King Jr.? Julian Bond? Muhammad Ali? Harry Belafonte? (Notice I'm leaving out Colin Powell, the key enabler of the most idiotic war in modern history. Sorry, Colin, but out of all those neocon knuckleheads, you knew better. You should have resigned in protest, made a big stink, and maybe, just maybe, this goddamn war wouldn't have happened.)

I'm lauding the kind of men who can read a room, who can smell a rat. These are the type of men who are willing to fight at any time, *but who only pick fights they can win.*

So yeah, Lou Gossett Jr.

It's 6 P.M., Monday afternoon. I'm all prepped for tomorrow's 11 A.M. with Oren Koules at Evolution. A fresh copy of the script tucked into my imitation leather briefcase along with the cast list, some other notes. Then the call comes. It's Chad Cole, Oren's assistant. "Hey, Rob, sorry but Oren's gotta postpone."

"That's cool," I say.

"How about later this week?" asks Chad.

This is good news because sometimes you'll get a call like this, and the assistant'll be like, "How about, um, let's see, oooh, how about next month?" And you'll be like, "How about you just take that headset and stick it up your ass because we both know your boss doesn't really give a shit and you know what, neither do I!" But do you say that? Of course you don't.

Because it's not just peons like me who are subjected to these bumps along Hollywood Highway. There are writers and directors and producers, people with substantial credits, who feel themselves being brushed off, brushed aside, thrown off-stride by the constant postponements, the unreturned calls, the waiting. The waiting. Is it disrespect? Of a sort.

But you've got to see it from the other side. As a lone writer facing endless hours of boredom and frustration, it's hard to understand why it's so difficult to make a simple meeting happen. But it's critical to put yourself in an executive's daily schedule. Oren Koules and Mark Burg aren't only producing movies and TV shows, their company is managing a whole slew of writers, directors, and actors. Which means they're constantly being hounded by their clients.

Koules probably receives more than 100 phone calls a day. And schedules an average eight, ten meetings. And then there are the working breakfasts, lunches, dinners . . . the coverage to read, the script notes to write. . . .

Anyway, Chad and I schedule for two days later, Thursday at 11. So I have to shift gears. How do I spend the rest of this day? First, I get to my emails and find this gem:

> Hi !
>
> I am a Polish basket agent. Can you give me e-mail adress to Aaron Goodwin ?
>
> Regards,
>
> Arkadiusz Brodzinski

Sure, Arkadiusz. No problem. And while you're at it, ask him why he and his brother aren't calling me back.

I wade through a bunch more emails, pay a few bills, then decide to spend a few minutes surfing the news. This turns up:

> WASHINGTON (AP)—The Army is investigating reports that several members of a reservist supply unit in Iraq refused to go on a convoy mission, the military said Friday. Relatives of the soldiers said the troops considered the mission too dangerous.

Wow. Now this is significant shit. Important enough for me to spend the next six hours learning more about instead of taking care of business. I come out the other end, dazed and confused. You know why? Because I thought we learned 30 years ago that wars like this are messy, complicated shit, and now, somehow, someway, we've let these assholes put our sons and daughters in the exact same position. It burns my ass.

Two days later, I'm finally back to my own shit—walking into the raw, sleek, chic offices of Evolution Entertainment on Highland Avenue (they own the building). There's a palpable buzz in the air. The buzz of green cash success. Between *Two and a Half Men* and now *Saw*, Evolution is riding high.

Saw is a low-budget horror movie ($1.2 million) which recently exploded at the box office and will go on to make over 50 million domestic. It was produced by Mark Burg and Oren Koules. Lions Gate

Entertainment immediately locked them into a nine-picture deal to deliver more horror. Blood and guts to fill the gutters.

I say hello to the receptionist, grab a *Variety*, and have a seat. A moment later, Oren Koules walks up.

He's got the muscular build, vise-grip handshake, and boyish good looks of the Canadian hockey player he once was. I follow him back to his corner office, where we sit and exchange thoughts about bringing along our young sons in sports. (Our shared belief—keep them versatile, mix up the sports and seasons, give them plenty of down time.) Then we get into the business at hand, *94 Feet*. Koules says he "loves" it. Thinks it should serve as the template for an entire series of low-budget sports movies. Hockey, football, baseball. Each movie a chronicle of just one game, start to finish. Track and field, that'd be a great one. A dual meet between two colleges—men and women competing.

Koules is unsure of the dual-ending gimmick. I lay out a convincing argument. It's a way of branding these movies. And it's true to the nature of sports—how the difference between agony and ecstasy can be as thin as a worn dime.

He buys it—people like conviction. We leave with the promise that we'll immediately chase ESPN as the logical outlet for these movies. "I'll call Bob," says Oren, meaning Bob as in Robert Iger, new head of Disney worldwide.

I hit the sidewalk with that wonderful bounce in my step. Hey, this shit works; the Evolution aura just rubbed off on me, and I'm defying gravity.

41

"YOU JUST WON A PWIZE"

Days pass. The phone doesn't ring. No magical email burps onto my screen. No word from Oren Koules on *94 Feet* and I'm beginning to worry. Plus I'm feeling despondent about *Hollywood Jock (the Movie)*. Shot down by Cube-Vision. Shot down by Brett Ratner's guy months ago. Shot down by *The Last Shot*, which bombed. Then John Carls calls. "I believe in you. I believe in a movie called *Hollywood Jock*. But it's not the three-pager you sent me. I don't care about two guys who've got dream jobs who want more. I don't care about the movie business this much. Hollywood should only be a backdrop. Get back to your stuff."

That helps. He's right. Forget the pitch, dump the three-pager. There's another movie lurking in the pages of this book. Something more along the lines of *Adaptation* maybe. Or *Swimming to Cambodia*. Or *Willie Wonka and the Chocolate Factory*.

I should have known weeks ago from Guma's reaction to the three-pager. I could tell over the phone that it didn't feel right to him, but still, I'm glad he didn't just shoot it down. He let me test the waters. It's a hard relationship between a producer and a writer or an agent and a writer. You shouldn't be too dismissive if you're not behind the eight-ball as the creator. These people have to hold their tongues sometimes. Let the writer take that first stab, then bring him around.

Hollywood Jock could make for a cool TV series. Maybe that's the angle I should pursue. But no rush. This one will come in good time.

Right smack dab in the middle of pondering the futility of the writing life I get a call from the owners of the Summer Pro League. Wondering why I disappeared so fast. "It didn't feel realistic that we could reach an acceptable price," I say.

"Let's try again," they say. "We'll send you some new numbers."

Aha. They're coming around. I make a few phone calls—I'm particularly concerned about the competing Vegas Summer League. Kurt Rambis calls back to say he'd be shocked if the Lakers ever left Southern California in the summer. So that's good. But then my old teammate, Geoff Petrie (who now runs the Kings), says they'll be heading to Vegas because his bosses, the Maloof brothers, own the Palms Casino there. He also cautions me to be careful, reminding me there's a lockout looming. Thanks, Geoff. Then Mark Cuban emails back that it's not guaranteed, but he wants the Mavericks in Long Beach next summer. Okay. That's cool. With all the uncertainty, this might be a great opportunity to buy the SPL at a reasonable price. We'll see what they come up with.

I email Clay Moser, who's the key guy to getting this deal done. Unfortunately, he's in China. Even though the touring team fell through, Clay landed himself a six-figure, six-month gig as head coach of the Changchung Tigers of the Chinese Basketball Association. Clay's fired up about buying the SPL. He'll be back by spring, just in time to kick the summer league into gear if we can make the purchase. Meanwhile, Clay's got his work cut out for him. He starts by shaving his arms. I'm not joking. His players are fascinated by the hair on his arms and have a habit of reaching out and stroking it during time-outs. What a world.

In the meantime, I receive an email from my new book editor at William Morrow:

Now that we finally have the contracts signed, just a quick note to say hello and see how things are coming along. If you need me for anything, don't hesitate to give a shout.

Be well.

Rob McMahon

I answer:

Hey, Rob

Thanks for the note. All goes well on this end. Plugging away. Generating new material every day. I'm jazzed about the book and wanna make sure you guys are ready with that second edition 'cause no way I'm gonna be happy with just 15k copies.

Best, Rob

Yo Rob—

Glad to hear all is well. What's going on with *Zulu Wave?* As for the book, if you deliver I'll deliver. Later . . .

I email back:

Zulu Wave—My cowriters and I have created a detailed outline and we're going to round robin our way through a draft by Christmas. This'll generate even more material for the book as we try to get it produced, Just heard from Mark Cuban—that the Mavericks will be back in SoCal for next year's summer pro league so that's good news

RR

End of email exchange. Things are feeling pretty shaky right now, even with the book. With Clay in China, Oren Koules not calling back . . . I shut down my computer and head upstairs to join my family.

Nothing like having a seven-year-old (he had a birthday) to keep it real. Andre still hasn't mastered his Rs. My wife (who's been a teacher and principal) says don't worry about it—these things self-correct in time. Still, I pull Andre aside and try a little amateur speech therapy.

"Try curling the sides of your tongue up so they touch your top teeth." He tries. I look inside his mouth. He wiggles his tongue at me.

"Try it."

He tries. Now say, "Ruh," I tell him.

"Wuh."

"Ruh."

"Wuh."

Yesterday we were hanging out and Andre decided to show off for his brother Cole and me by making a crank call to our friends, the Starbards.

Lana the mom picks up. "This is Mr. Baskin-Wobbins calling," says Andre as if this is the funniest moment in history. "What ice cweam is yaw favowit?"

Lana decides to play it straight. "Uh, chocolate chip."

"Thank you," says Andre. "You just won a pwize." And he hangs up roaring with laughter. Days later he runs into Lana and her two teenage boys and Lana says, "Andre, we got a call from Mr. Baskin-Robbins and we won a prize."

Andre laughs. "That was me! I punked you!"

42

$800,000 A YEAR

George, Mark, and I spend five straight days hammering out a scene-by-scene beat sheet for *Zulu Wave*. This is probably the most enjoyable part of writing a screenplay. It's the time you spend trying to see, then describe your movie. What happens? What happens next?

We're on the phone back and forth, sharing the vision until it's unanimous, then charting the thing out in 80 scenes or so. After worrying about the dynamics of three guys writing a screenplay, I'm beginning to see the upside. There's an instant mediator when disagreements arise. Majority rules.

The beat sheet (it can also be called a step outline) is invaluable. We used to do this stuff on three-by-five cards. Spread 'em out on the bed. Tack 'em to a wall. Each card bare bones:

> EXT.— NIGHT—WHITES-ONLY BEACH.
> Kwezi surfs alone. Cops arrive.

This practice is especially valuable in tracking secondary characters. "Look, she was here for three straight scenes, then she falls off the map until act three. We gotta bring her back sooner."

Okay, we're in agreement on the movie we're writing. So Mark volunteers to knock out the first five to eight pages. And I'm thinking, You go, boy. Hey, this could be perfect for someone like me who's got so many irons in the fire. Give a burst of work, then back off as your partners grab their picks and shovels.

There's suddenly renewed heat on the Summer Pro League possibility. I've got to pay attention here. I spend a couple days shoring up my potential investors. They swear they're good for the dough. Then my old friend Nigel Miguel calls to say he's got someone else expressing interest—an African-American music video director named Phillip Atwell. He and his guys are interested in learning more. They'd also want to keep the ownership group tight. So I talk it through with Nigel, hang up and Google Phillip Atwell, and this is what I find:

SUGE KNIGHT CAUSES FRICTION AT 50 CENT VIDEO SHOOT

> The infamous Suge Knight came uninvitingly to 50 Cent's "In Da Club" closed video set with 13-15 Mexican gangbangers, trying to stir fear in the set unsuccessfully. 50 Cent was not having it and stood ground with G-Unit and Shadyville while camera crews were horrified. Suge Knight just came and stood there for five minutes with his Mexican crew, smoking a cigar, then left without saying a word. FBI and detectives are currently watching every move Suge Knight makes; but we wonder why they weren't around the set?

This "In Da Club" video will hit MTV and BET in a couple of weeks. It features Dr. Dre and Eminem as two doctors creating the $6 million "negroe = 50Cent." Directed by Phillip Atwell.

So much for Phillip Atwell. This is the kind of crowd he's running with? Who needs it?

If it's unfathomable to you how guys who've beaten their way out of the ghetto to get incredibly rich can so quickly resort to violence, thereby risking everything they've gained, you're not alone. But me, I get it. These guys are dumb. If you don't believe me, ask Chris Rock.

Hey, here's an idea: Why don't I just call Suge Knight and see if he wants to own the SPL. That's a good one—Suge Knight and David Stern singing "Tiptoe Through the Tulips."

I shore up our other potential investors, then work up a counteroffer in an email. I send it to Clay Moser in China, plus a couple other advisers I have. Clay emails back: Sounds good, but shouldn't this be in more of a letter-of-intent style, with a counteroffer date?

This throws me. I was wanting to keep it informal. Avoid the legalese. I Google "Letter of Intent" to learn that any offer, to buy anything, is in certain circumstances considered a legally binding contract. Whaaat? You mean two guys can't throw some numbers back and forth without getting the lawyers involved? Fucking lawyers. This is an uncertain rant here, because I personally know a lot of lawyers, and I like them. They're smart guys, truly engaged with the law. But there are a whole slew of them that I gotta lump in with the stockbrokers and bankers and agents and tax accountants and all the other brainy greedsters whose main purpose in life is to create huge gobs of complex, convoluted language that only they can decipher, enabling them to skim huge gobs of wealth which was created by people who actually work for a living.

So, a message to all you middlemen (you know who you are): You're taking way too big a portion of the pie of wealth being created by people who actually do real things. Like clean houses and make brake linings. Deliver packages, teach our children, grow food, make shoes. Without people doing all this stuff, you'd have nothing to broker.

I've got an entertainment lawyer friend in L.A. He works pretty hard. He's got some big clients, so he spends a lot of time jawing back and forth with studio and record company lawyers. Hammering out deals.

They all know each other and pretty much get along and after racking up the usual number of billing hours, guess what? Wow, whoop dee doo—they finally reach agreement.

For this, my friend receives a base salary of $800,000 a year—$800,000! Plus he can receive bonuses worth another couple hundred thousand on top of that.

Whose money is this?! It *was* yours. Until you dropped that 80 bucks for a shitty seat at a Beyoncé concert. And Ticketmaster grabbed another $8.50 for a service that couldn't have cost them 50 cents. (Don't get me started on Ticketmaster.)

Or until you tried to put a few thousand measly bucks into an IPO, only to learn that a select group of bankers had first dibs.

Message to W and his cronies who want the American people to invest in their own futures. Get a fucking clue: We're all broke! Don't you get it—Americans are broker than broke. Credit card debt, tuition debt, health care debt. There's nothing to invest!

But my lawyer pal is sure gonna take advantage of all these new rules.

Okay, enough. (Too much, I'm sure my editor will say.)

Anyway, I finally send off a counteroffer to the SPL people—stressing that it's simply an informal exchange of possibilities and in no way legally binding. Then I also throw in this bit of legalese I lifted from somewhere: "Naturally, any deal will include the usual representations and warranties and pertinent noncompete and/or nonsolicitation clauses. The offer is also conditioned on the completion of satisfactory and customary due diligence acceptable to both parties." So take that.

I reach Friday, feeling disgruntled about my workweek. And feeling worried that the weeks are speeding by. Forty-two gone and counting. Plus I'm doing too much business bullshit. So I hit *Zulu Wave* hard, and in two three-hour bursts, I write six good pages. The first time we see Kwezi on a surfboard. Introduction to his buddies, Solly and Mussa. All right. This is feeling good now. This story wants to tell itself.

So I'm about to close up shop for the weekend. Spend some family time. But Oren Koules is still gnawing at me. Jesus, it's been over two weeks now. During which time *Saw* goes through the roof at the box office—so maybe that's it. I've called. I've emailed. No response. So I send off another email:

> Oren—
>
> Let's go, man. I know you're raking in the dough, but I want to get moving on *94 Feet* and the others. Have you spoken with ESPN?
>
> Best, Rob Ryder

43

"CANNED HEAT"

Bright and early Monday morning I hit my basement office. I've got *Zulu* on my mind. But then I receive a call from George. He and Mark just got a great response from those hotshot young English producers on another project. The producers want to pay them to deliver a script, and they want it fast. George has to put *Zulu Wave* on hold. He hopes I understand. I understand. Money talks. I'd have done the same thing. But still, it hurts. We were just getting up to speed there.

What now? I feel the panic seeping in. I'm running out of time. Maybe it's best that *Zulu* drops out of the equation. It's so easy to lose yourself in a spec writing project. Keeping your nose to that spec script grindstone keeps you blind to that yawning abyss known as financial ruin just downriver. I need to be quick on my feet here. I need to make some money.

An email comes through from Scott Matthews—an African-American producer-promoter-entrepreneur whom I met through my column. He's been working with black uber-agent Aaron Goodwin on a project called *Hook*. Hook's a real-life streetballer out of Oakland who ran with Gary Payton and could jump over Volkswagens but ended up in the slammer for five years for larceny. Then he got out and cleaned up and is now a community activist and a pretty cool cat.

Goodwin reps Hook. He thinks there's a book and a movie there. He's asked Scott Matthews to oversee the project. Scott says they need a treatment. Wondering if I'm interested. I email him back:

> Hey, Scott—
>
> I'm interested. I think it's a terrific story. I can jump on it right away. But here's what I want (and it's minimal). $1,000 to write a ten-page treatment (plus a few hundred bucks in expenses). And I'll want a one-page agreement that I've got the rights to write the first draft of the screenplay. I can have the treatment done in two weeks. I'm also willing to use my considerable connections in Hollywood (at ESPN but elsewhere as well) to get the movie made. Mark Cuban's 2929 might be interested. I can talk to Mark about it once there's a treatment. Plus there are others—Broderick Johnson at Alcon Entertainment, Andrew Jameson at Maloof Television, etc.
>
> Let me know.
>
> Best, Rob

Scott calls. We talk. He says he'll run it by Aaron Goodwin.

Okay, so that's something. Maybe. Another ball in the air. Regarding *94 Feet*, I've pretty much written off Oren Koules by now. But I'd hate to see it just die like that. Unfortunately, some movies have a shooting window. *Snow Dogs* you're not going to be shooting in the summertime; *94 Feet* you are—because that's when the arenas are sitting empty and all the ballplayers are available. I can feel the requisite preproduction weeks closing in on that one. Another month slides by and *94 Feet* won't be getting made this summer. It's painful.

I consider calling the guys at ESPN.com and suggesting another article, but about what? It's dangerous when you find yourself losing interest in the things that used to rev you up. Or is it? Hey, maybe there's an article here: "10 Best Reasons Not to Give a Shit About Sports Anymore." Nah, they'll never buy it. Besides, for a measly five hundred bucks, I'd rather go to the credit cards. Although maybe Salon.com will e it away in my spiffy-jiffy FUTURE PROJECTS file.

the afternoon staring dully at my personal library shelf. A couple years ago, after a near-death encounter with a blood clot, I decided to finally catalogue all the work I've done in the past 30 years. Jesus, what a task. It's like carving headstones for a bunch of rotting corpses, some of which you don't even recognize anymore. But I pretty much got it done. I've got two copies of every script I've ever written—one in a labeled manila folder, one in a three-ring binder with the title spelled out on the spine.

Okay, here's a tip: If you have any aspirations at all about being a writer, if there's anything at all you can take away from this book, here's the best advice I can give you—print everything out on three-hole punch paper and stick it in a binder.

A title catches my eye: *Are We There Yet?* No, this isn't Ice Cube's new movie, it's a script I wrote years ago which still has a lot of promise. I groan. Now I have to find a new title. This has happened to me before. *The Sky Is Falling* was my first decent Hollywood deal. A few years later an Australian thriller came out bearing the same title. There's no protection of titles; you can't register them like a script—it's all about who gets there first. But that fast, I come up with a new title, and I'm jazzed about it. *Canned Heat*.

See, *Canned Heat* (formerly known as *Are We There Yet?*) was another of those oh-so-close disappointments. It's a movie about a group of failed yuppie New Yorkers and their dysfunctional children who chuck it all and move to a house in the Catskills. And being a comedy, it all turns to shit in about three weeks.

Canned Heat is one of those that I've put a lot of time in on. Over several years, I've done two readings with talented actors in front of live audiences. I've rewritten and rewritten. I've met twice with Beau Bridges, who really wants to get it made, and also with the talented director Peter Werner. I've even pounded out a novelization, which is how I met my book agent, Matthew Guma. It's rich, sexy, funny material speaking to an entire generation (and their kids) for whom the shit is hitting the fan.

(Interestingly, as the years go by, the story only becomes more relevant, as more and more broke yuppies reach retirement age.)

But I haven't been able to get it made. As for the novel, Guma says he can sell it if I put in the time on a rewrite. I'll get to it. But more immediately, maybe it doesn't belong on the big screen. Maybe it'd make a good series. I call my pal J.B. White.

J.B. and I spend a few minutes getting caught up. It seems that after being stuck in limbo for a few weeks, J.B.'s movie *The Twelve Days of Christmas Eve* has been shot and is in the can. Power to him. That's gotta be a good feeling. And that's the thing about television. Six months ago, his project was a bare-bones idea. And that fast it was pitched (for free), written (for hire), and produced (for back-end fee). Bing, bang, boom. Television.

J.B.'s been to both my readings of "Canned Heat" and even been so kind to have read the novel (which I embarrassingly forgot. Jeez—it's one thing to make friends suffer like that, but then to have forgotten the sacrifice they made . . . !??)

Anyway, I talk through my conception of *Canned Heat* with him. J.B. likes the idea. Says he'd be willing to read it if I come up with a pilot script.

So I spend the afternoon rereading *Canned Heat*. Aha—this 120-page script could easily be broken in half and rewritten as an hour-long pilot for television. So I get to work. How? First, by finding several pilot scripts on the Internet and printing them out, just to get the form down.

Three days later, I've got a 55-page spec television pilot. This is not normally the way TV series get sold. It's usually through established TV guys pitching ideas to executives in meetings set up by their agents and managers. But the TV world is topsy-turvy right now. More open to reviewing projects coming through less established routes.

J.B. likes it. Advises a few changes. I tell him I'm gonna send it off to David Madden at Fox Television Pictures. J.B. knows David—thinks that's a great idea.

I take a deep breath and ask, "How do you feel about showing it to your agents at William Morris?"

This is a loaded question. Decades ago, agents seemed to come as cheap as writers. That equation has changed. So writers who are fortunate enough to have reputable agents are not so cavalier about the chits they call in for fellow writer friends who don't have agents. If J.B. decides to say yes, it's with the understanding that if the pilot somehow, miraculously goes to series, he'll be first on the list to get called to help write and produce.

Instead of yes, he says, "Let me send it to my brother-in-law first. Get his reaction."

Fair enough. David Babcock (the bro-in-law) is a working TV director-producer who's had a few nice runs. But clearly, like most, he'd be the last to say he's sitting comfortably. No one is. Except for guys like John Wells and that *Ally McBeal* guy who's so famous I've obliterated his name from my memory.

J.B. tells me about how Babcock has been developing a personal favorite of his for years. A TV series (it's based on a slim French novel) about a perfectly normal guy—dresses well, married with kids—who has only one aberration: He was born with, and still has, the head of a dog.

Babcock's been bringing this idea along for over 15 years. When he first pitched it to his agents, one of them immediately responded, "David, I think it's better for all concerned that we pretend you never told us that idea."

Now, years later, in this new era of edgier, over-the-top shows, his agents are all over it. You can't help but wish him well.

So *Canned Heat* the pilot gets sent off for first reactions from two pros.

Now what?

Usually to get away from my miserable self, I surf around the world news. Think you got it bad, try Darfur! Today, I try to stay sane. So I go off to buy groceries. And I run into my friend Bill Couturie. And Bill Couturie looks like he's been hit by a truck—his face haggard, his body slouched, his eyes dull. There's a reason for this. See, way back in the late '80s, Bill Couturie won an Oscar for best documentary, *Dear America: Letters Home from Vietnam*. And here he is, years later, in the middle of editing his latest doc, this time for HBO—*Last Letters Home: Voices of American Troops from the Battlefields of Iraq*.

For the past several months, Couturie and his tight documentary crew have been knocking on the doors of families across America who've lost sons or daughters in Iraq. Then they've stepped inside, set up cameras and mikes, and recorded these mothers, fathers, husbands, and wives as they read the last letter their loved one sent from Iraq before being killed.

Can you imagine? Now that right there'll take you straight outta the whiny, self-pitying morass you've blundered your sorry ass into. No wonder Couturie's looking worse for wear. Sitting in the room as parents bawl their eyes out. As fathers cannot even bring themselves to speak for fear of breaking down, and so let their oh-so-brave wives bear the burden. But the father needing to sit there. To at least be on-camera. To bear witness. To summon that incredible bravery to face the death of your child.

What the fuck are we doing in Iraq?

I ask Bill. "What the fuck are we doing in Iraq?"

He stares at me, his eyes like stones. "I don't know," he answers. "Losing another war?"

At home that night, on the California king-size bed which doubles as a trampoline for the family. Playing "Panthers and Jaguars" with Cole and Andre. The three of us on all fours, circling each other, snarling, eyes growing wilder, until finally we all pounce at once—wrestling, tickling, screaming . . .

My boys. My boys who will never fight in a stupid war.

Then a call comes from George. He talked to Mark. They said to each other, "What are we crazy not to keep writing *Zulu Wave* on spec?" So they're back. They want to resume writing immediately.

44

THE SUPERNATURAL REDNECK VERSION

First thing Monday morning I find an email from Mark—he's made one of those sizzling 20-page runs on *Zulu Wave* over the weekend. I open the "Final Draft" attachment. It's good stuff. If we push, we can have a script in a few short weeks.

Plus, the three-writer arrangement still gives me time to deal with all the other shit going on. Like the possible treatment I'm gonna write for *Hook*. But then this arrives—Aaron Goodwin's response upon learning that I wanted to be paid $1,000 to write a 10-page treatment:

> Scott,
>
> Tell him to write the treatment for free if he wants to be involved, and if the project moves forward, we will include him, otherwise, I will get someone else to do it.
>
> AG

Hey, that's the word from the great Aaron Goodwin. The man who represents 20-year-old LeBron James, who's rapidly turning himself into the best basketball player in the world. Hey, Aaron, you're cutting deals worth tens of millions of dollars. God forbid you try to loosen up a mere 1,000 bucks for some lowly writer when you know there's some other lowly writer willing to do it for free.

So much for *Hook*. Boo hoo hoo.

What next? George is making the next run on *Zulu*, so I clock some more hours working this Summer Pro League purchase. Wouldn't it be a trip if this is what comes through? We've built enough capital into the bizplan to put me on a full-time salary at least through the first summer season. That could be great. Talk about having your life pivot on a dime. And the offers and counteroffers are actually growing closer. I remember what Dennis Ross, the Mideast peace negotiator, once said: "The more time and effort both sides involve in the negotiation, the better chances you have of reaching agreement." I'm putting in the time. If this comes through, this could be just the financial ticket I've been looking for—a piece of an ongoing seasonal business which could be worth tens of thousands of dollars a year for me personally. Plus I wouldn't have to write anymore. (Thank God, you're all thinking.)

I saw the wonderful writer Ricky Bragg on C-Span the other day. An audience member asked, "Do you like writing?" A long, long pause—a pause so long it elicited ripples of laughter. The answer finally came—"No. I like *having* written."

And I like having written *Canned Heat*. Because I finally get feedback from both David Madden and David Babcock, and it's positive. But they both think it should be kicked into a higher gear. Babcock suggests I heighten the tension between the New Yorkers and the locals, bring some *Deliverance* to it. Hmmm, that's a possibility. Madden responds that a supernatural element might put it over the top. You can't be too outrageous these days. Hmmmm, maybe the old farmhouse could be haunted.

But you know what? I hate anything to do with the supernatural. Mostly because it's all fake. Secondly, if it's done well, it scares the crap out of me, and I've gotten enough to fear in the natural world. But hey, when Madden talks, I listen.

Madden ends our conversation by saying there's a show called *Point*

Pleasant debuting soon on ABC and that *Canned Heat* will inevitably be compared to that. I hang up and Google *Point Pleasant*.

A mysterious teenage girl turns up in a New Jersey resort town and starts acting witchy. Say what? But, hey—you want witchy, I can give you witchy. So I spend the next couple days rewriting scenes to reflect this newfound sensibility and send the pilot script off into the etherworld once again.

It's Friday afternoon. Through all the other flailing away, there's one project that's really keeping me hooked in: *Zulu Wave*. Against all odds, George, Mark, and I are cranking out more and more pages, sticking to the beat sheet, rewriting each other so all the voices blend into one. It's working. And it feels great at this stage—you know why? Because it's still under our control. It's still being created. We're not waiting for any phone calls.

I'm actually feeling elated. Working my ass off on the one project that's keeping my hopes alive. Come 6 P.M., I wrap up, grab my 12-year-old, Cole, and a soccer ball and we head off to a local elementary school field. It's times like this that a dad lives for. Balance. The ability to actually help your kid through some stuff when years ago your dad might not have been there.

I once caught this piece of advice, and it stuck: When you talk to boys, make sure you're doing something else at the same time—playing catch, kicking a ball around, baking brownies. They'll be much more inclined to share—won't have that eye-contact thing to throw them off-stride.

I'm thinking how it's probably gonna be that basketball-soccer dilemma. Cole's been splitting his time pretty evenly between the two sports, but soon he'll have to choose. Though I'm a hoops guy through and through, I'm gently guiding him toward soccer. Why? You get to play outside, you don't have nearly as many assholes in the stands screaming at you, and Cole, being a white boy, has a much better chance at a college scholarship pursuing soccer than hoops.

But it's his call. And if he decides to drop them both, I can live with that too, just so long as he doesn't turn into some Goth/slacker/junkie/malcontent or something.

But that's not what he needs to talk about. Instead, Cole relates how his friend Quinn got sucker-punched in middle school the other day. Quinn and a "buddy" were fake-sparring when the other kid lost it and threw a hard right hand to Quinn's nose. Quinn and his assailant both got suspended. Quinn, being the reasonable, savvy kid he is, decided to let it go. But that's not good enough for Cole. He wants to challenge the bully to a fight. I look over at that sweet 12-year-old face, that changeling body, that $4,200 worth of orthodontic work staring back at me.

What am I supposed to say? (Sorry, I just don't have the automatic politically correct response here.) Instead I ask, "Think you can take him?"

"I know I can take him," answers Cole, his eyes locked in on mine.

This isn't an easy one. See, being a guy, whether it's at age 11 or 38 or 73, there's a very specific type of self-loathing that kicks in if you back away from a fight. We've all been there. Every one of us.

My latest experience was 14 years ago. In the small gym at the Hollywood Y. The usual suspects forming teams; a couple of flashy new guys showing up. One guy was Stoney Jackson—a young black actor with a hard body and Jheri curl. I vaguely knew who he was—he'd appeared in Walter Hill's *Trespass* and he'd been popping up on TV pretty regularly.

Anyway, the games at the Y could get pretty physical, and I soon found myself under the basket with Stoney Jackson, boxing each other out for a rebound. I played a simple trick on him—I hooked him a bit, and leaned into him hard. But when he pushed back I stepped to the side, and he fell. Then he popped back up to his feet—everybody hooting at him—and punched me with a left hook to my right cheek. I was stunned. I felt a gush of blood inside my mouth. Then he backed off, both hands raised. "Come on, motherfucker," he said.

I thought for a moment. I'm a board member of the Hollywood-Wilshire YMCA. I'm chair of the P.E. Committee. One of our biggest issues is fights in the small gym. I can't fight this guy. So instead, I find myself saying, "You know what? You're outta here."

"Say what?"

"You're outta here. You can't punch somebody on this court and get away with it."

"You threw me down," Stoney Jackson spat back.

"I didn't throw you down. I stopped leaning on you and you fell." The words were coming okay, but I could feel my jaw quivering, my anger barely under control.

"You can't kick me outta here," he said.

I look over at the group of regular guys—K.B., Rev, Mark—and I can see they're with me on this. Suddenly somebody says, "Let's just call the cops, man. The dude assaulted you."

This gets Stoney's attention.

"No one's calling the cops," I say.

By now the P.E. director (whom I helped hire) is in the gym, sussing things out. In short order, he fills out a bad-conduct card banning Stoney Jackson from the facility, gets him to sign it, and shows him the door. And I play on as if nothing happened.

But I go home feeling like a total schmuck. A chickenshit momma's boy. The feeling lasts for days. Even now, the memory stings. I should have hit him back. I should have hit him back, then had him kicked out.

Consider the world we're leaving our children. Look at the developing chaos. You don't think it pays to teach our sons (and daughters) to physically defend themselves? I want my kids to project this aura: I'm a reasonable person, but I will physically take you on if I have to. Is that such a bad thing? You're raising a child in this world, you'd better be raising a warrior. A defender.

But this is a 12-year-old boy we're talking about here. How clear is his decision-making? You can just see the testosterone pumping through his bloodstream. It occurs to me that maybe he's just thinking it'd feel good to beat the shit out of someone and now he's got the excuse to do it. Ah, how quickly the pendulum swings from Defender of the Innocent to Nazi Youth.

"You know what I think?" I ask.

"What?" he asks.

"I think it's not your fight."

45

DERANGED CHRISTIANS AND OTHER WMDS

Here's the deal about celebrities. Celebrities exist so we can vent all our (perfectly natural) mean-spirited, small-minded gossip on them instead of on our friends and neighbors. It's way healthier. Hey, wanna take a shot at Ben Affleck? Feel free. He asked for it. When the guy behind the counter called out, "Who wants to be a celebrity?" Ben raised his hand. So take out all that gnarly, putrid, pea-brained anger and resentment on Ben. Better still, share it all with your friends and neighbors—and watch them pile on, free of any consequence for spewing their meanest thoughts. You'll all feel better in a matter of minutes.

Why am I talking about this? Because I just read an article about a man in London smashing a Nativity scene with a hammer. Why—because he was outraged that the Virgin Mary and Joseph were perfect likenesses of David Beckham and that Spice Girl he married, Posh. Posh Spice as the Virgin Mary.

Wow. You gotta ask yourself, What were they thinking? A bit of that lowbrow British sense of fun, I guess. One headline read—"Beckham Nativity Scene Damaged by Deranged Christian."

My jaw dropped as I went on to read that actors Hugh Grant and Samuel L. Jackson portrayed shepherds (say it ain't so, Sam!) and that

George W. Bush, Tony Blair, and Queen Elizabeth's husband portrayed the Three Wise Men.

I swear to God. I'm not making this up. I Google a couple more sources just to make sure. Sure as shit. Can you spot the deranged Christian in this picture? You can just hear W (the Wise) whispering to Tony, "Hey, boy toy, can you cut me a couple lines of that frankincense you're carrying?"

And you can just hear Beckham, playing the cuckolded husband Joseph, making his case: "Hey, maybe some other guy did screw my wife, okay? But at least it was God!"

Don't I have work to do? A career to salvage?

Sure, and come the start of another week, it's all *Zulu* all the time. Or almost. We've got this script cooking. Gobs of pages spewing through the Internet pipes. We write, we rewrite. We stay in sync. And suddenly we've got 60 pages. Seventy. Seventy-five. We're closing in here. Pushing each other. Discussing, making our arguments, deciding. And moving on.

And the beauty of it—there's time in between. A third the labor. (Of course, that means a third the financial reward if we're blessed enough to sell the thing.)

By now I've also been clocking some serious hours—week after week—pursuing the Summer Pro League. I've spared you most of the mundane details. This is the kind of sharp left turn my life is crying out for. A job I can do and do well. A bottom line. An opportunity to build a small valuable business to ride into the sunset. Or sell.

So I continue the negotiation. The owners are selling me on potential, while I'm trying to buy the reality. I'm suddenly concerned about the "what if" factor. What if we buy the thing, and there's a lockout, and the Lakers never show? The answer—we take a financial bath. Are our investors willing to face that possibility? Can they weather that storm and move on to the next season? I email Clay Moser in China—what do we do in case of a lockout? "We make lemonade from lemons," answers Clay. Oh, great. Thanks a lot. Clay knows as well as I do that in a league like this, you make lemonade from Lakers. And suddenly the pro forma spreadsheet I've created needs to be amended to account for this NBA lockout scenario.

Everything is hard. When you're starting a business, you've got to weigh the "what ifs." Just ask all the guys who lost their shirts in the Arena Football League. Ask all those knuckleheads purchasing ABA franchises for 10 grand apiece. You've got to have enough dough to weather bad box office as you build franchise value and finally turn the corner into profitability. You've got to plan for the worst-case scenario.

I turn my attention to *Canned Heat*. Still no reaction from David Madden at Fox to the supernatural redneck version. Hey, there's a TV show right there: *Supernatural Redneck*. Or at least it's a country-western song. I call my buddy J.B. to see if he's spoken with his bro-in-law, David Babcock.

"No, David's caught up in a not-for-airing pilot."

"What's that?" I ask.

"The networks finance a bare-bones video production of the pilot script. They actually shoot the whole thing, sets and everything ... then they decide whether to commit to the actual pilot."

"What do they spend on these things?"

"Fifty, sixty grand. It's down and dirty."

"You're kidding," I say (remembering that companies are spending three, four million dollars and up to shoot an actual pilot).

"Anyway, David's tied up."

"Okay," I say. "He's been real helpful so far." A pregnant pause—I consider asking J.B. how he'd feel about showing it to his William Morris agents, but I hold my tongue. We'll wait to hear from Madden.

J.B. tells me his TV movie, *The 12 Days of Christmas Eve*, is going to premiere next Tuesday on USA. They've got to hit a 3.0 in the ratings for it to be considered a respectable showing. A 3.0 translates into fewer than 3 million viewers. God, how times have changed with the fracturing of the dial. J.B. White once wrote a miniseries called *The Beast* which was watched by 32 million Americans in one two-hour period. So he'll be going from 32 million to 3 million (and he'll consider himself lucky if he hits it).

I still can't get Oren Koules out of my head. My last hope that he's simply too swamped to respond has disappeared, replaced by a brooding anger—Man, you could've at least had Chad call me back: "Oren's swamped, he apologizes." Even that would have been enough. But no, you can't even afford me that tiny modicum of respect.

No, you're too busy counting the dough you're making off *Saw* and the eight other horror films you're contracted to produce for Lions Gate. It turns out my son Cole actually saw *Saw*. This is what he said.

"It was kinda weird, you know, how they had this woman and she's got this bear trap on her head and it's like slowly expanding and tearing her head apart, but she needs this key to unlock herself, but like the key is in her husband's chest and she has to cut her husband open with a knife to pull out the key and save her own life. It was weird."

I bet it was. All the way to the bank. It makes it easier somehow —if Oren Koules is happier producing sick, twisted stories in a world already overflowing with too much real-life guts and gore instead of movies with some socially redeeming value, so be it. Me, I'm committed to the integrity of *94 Feet* and to the humanity of *Zulu Wave*.

So. Once again, *94 Feet* will sit out another season. It hurts. But once again, my now seven-year-old, Andre, pulls me out of it. We're hanging out by the pool, and Andre's got that look in his eye—there's something he wants. And it's gotta be something big, 'cause he's taking his time, working up his opening line. I run through the usual litany in my head—it's gonna be either a 50cc dirt bike, a gas-powered air-soft shotgun, or a monkey. How do you say no to any of these wonderful additions to our family? Let me count the ways. But it hurts to say no, particularly when your kid wants it so bad and makes the effort to be creative in how he asks for it.

I grimace as Andre screws his face up, reaches out, puts his hand on my arm, and says, "Dad, how long have we known each other?"

46

"JUST WHERE HAVE YOU BEEN?"

As if things aren't tough enough as they are . . . my HP Pavilion 750n desktop computer is slowly grinding to a halt. Ridden with enough viruses and spyware for even the most jaded techie to wince and ask, "Just where have you been?"

I plead the Fifth. It seems that I haven't been updating my antivirus software religiously (and that's what it takes). And that's what I get for sharing my work computer with my boys (who careen through the World Wide Web with total abandon). Not to mention the fact that my friend Bill Turner was just recently caught logging onto my computer and checking his emails without asking me first. Jeez Louise! What do you think you're doing, Bill? I don't care if you are a super-cool art gallery owner who can throw a football 70 yards and talk politics until 4 A.M. Don't you understand that you just committed the most vile trespass of our day?! I mean, I'd rather see you stepping out of the shower with my wife than see you logging onto my computer without permission. Uh, actually, I take that back. But you get what I mean. Leave my machine alone! It's very sensitive. It's laden with viruses. It's doing its best to hang in with me as we finish *Zulu Wave*. One more random click could spell its death knell. So, in one sweeping decree—all members of the human race (especially family members) are banned from using my computer until this script is finished.

"All right," they say. "Stop being so crazy."

"Stop being so crazy? You want me to stop being so crazy? Then stay off my machine." You see there's a particular emotion—brand-new to the human race—that's caused by computer failure of any and every sort. It's a paralyzing combination of rage and utter impotence. It's a psychological disease (I'm sure they've got a name for it by now). And God knows I've got it in spades.

"Okay, Dad," answers Andre. "We undastand whey-uh yaw comin' fwom."

I call my local techie, who talks me through a temporary cleansing (at 80 bucks an hour, mind you). Okay, it's running well enough to blast through the end here. A couple days later, Mark, George, and I have our pages done. We dutifully email them to each other and paste them together in a master draft. It's running about 110 pages. We separately read through what we've written. We like what we read. We decide to get some fast feedback, first from our wives. (Normally a dangerous proposition, but June, Liz, and Andrea have been fully apprised of the process and have given us full approval to write this story on spec. And to a woman, they like the script.)

Their notes are minor—clarifications, mainly.

All right. That feels great. But we all decide to take a few days. I send the first draft to Bill Couturie. He's tougher. He thinks the story is driven too much by the action—character is getting lost. We discuss it with our wives—yeah, sure, there's room to go deeper.

So we plot out a rewrite. And just as we're about to plunge into it, my computer finally dies. I mean you can see the smoke coming out of its speakers. And if it could have written its own gravestone epitaph it would have been something accusatory, like "I told you I was sick."

Okay, that's it. We're switching to Mac. No more of this bullshit. I don't care what it costs. Andrea and the boys are totally with me on this. I go to the online Apple store and lo and behold, they're offering not only new computers but reasonable-rate credit cards. Wow. Not only do we get a new Mac—but it's free! And on top of that, we've just signed on for an additional $6,000 of spending power! For free!

So we all pack into the minivan for a trip to Apple Store at the Grove on Third Street in the Fairfax district. In the meantime, I've got my local techie (at $80 an hour) salvaging everything he can from my old deep-fried hard drive and putting it on a couple of disks for me.

A few hours later we're home. Our new machine is up and running and Mark, George, and I race through a rewrite of *Zulu*. This thing is feeling solid. Hey, we've got a movie here!

You see, you go through these waves of euphoria when you finish a good one. You feel like screaming and shouting, "Hey, we got a movie here! Getch your red-hot movie!" You feel like you've written the only script in the universe. It's the only one! This fucker is sizzling, baby! But then, if you've been around the block, you go into the dark-side conviction, "If you can't see this one, then I'm dead meat. If you don't get that this script deserves to get made, and for all the right reasons—profit, prestige, and socially redeeming value—then I'll slice my fucking balls off and crawl out into the desert and die!"

They gotta get it. They got to.

47

LOST SOULS AND DROOPING DERMATOLOGY

Okay, enough dredging this stuff up from the past. No more of the boo hoo hoo. I'm about out of time. So from here to the end, it's all business. I promise. First I've got to convince my beautiful wife that I'm this close to cracking something here.

It's been tougher than tough on her. Since she walked away from her job (months ago now), the terrific period of rest and recuperation that she'd imagined—just kicking back, doing homework with the little guy, baking cookies—has been tainted by the dwindling bank account and the growing evidence that I'm probably not gonna make that miraculous score that'll forever free us from month-to-month financial panic.

All along, I've been prepping her about how I might need an extension. Throwing these little bones at her—little glimmers of hope that'll keep her from pulling the plug. But it hurt when we had to start going into the credit cards to pay the bills.

Still, I showed her the pension statement from the Writers Guild of America. How my last substantial deal (at Paramount) had vested me in the WGA pension fund. And how every WGA deal from now on (no matter how small) will add real dollars to the monthly stipend I'm guaranteed (as long as it doesn't go belly-up like the United Airlines pension fund did).

Okay, that's something. But I'm years away from collecting some meager pension. And we've got bills to pay. Suddenly, out of the blue, a friend emails her about a productivity consultant company in dire need of a polished, articulate multitasker. On almost a lark, she applies, and two days later, she's hired and finds herself at her new desk. Whoa. How did that happen?

It's a shock to the system when you take months off from work, then are suddenly reimmersed in it. The guilt ran deep as I saw her struggling to regain her sea legs. The people at this new company work hard. They hired the right person (and gave her a substantial raise after her first day on the job). But it wasn't easy for her.

When she drags herself home after her first day, I throw her another bone. I tell her that I've been contacted by Glenn Stout, the editor of *Best Sports Writing of the Year*, about submitting an old ESPN.com column for the next edition.

It isn't exactly a lie. He did contact me.

"Wow, how'd they track you down?" she asks.

"Probably through Michael Knisley," I say (now lying through my teeth; Knisley was my old editor at ESPN.com). Look, it's like this—technically, Glenn Stout had contacted me. But only after I had contacted him. I came across his name when surfing around for information about the late great sportswriter Dick Schaap. Somehow I turned up Stout's email address and I asked him the protocol for submitting a piece. His reply: Print it out in hard copy and send it to me at the address below. Be sure to include name and date of appearance.

(Hey, thanks for the encouraging tone, Glenn.)

At least this sportswriting bone makes sense to her. She saw how I'd been making $500 a week at ESPN.com for only a couple days' work. I guess she did the math—$500 for two days' work . . . that'd be $1,250 a week for five days' work. That's 5K a month, and that's respectable. That'd almost cover the monthly nut. You add on her new salary and we'd finally be banking some dough for the future.

So forget my latest bloody mess of a stab at sportswriting for ESPN—let me learn something from a pro. I find an interview with Dick Schaap by Jenna Glatzer on AbsoluteWrite.com. Glatzer asks: "Any further advice for aspiring sportswriters?" (That'd be me and a couple million other masochistic morons.)

Schaap's answer: "My advice to an aspiring sportswriter is to read the best, from Shakespeare to Kornheiser." (Sure he doesn't mean Jason Alexander playing Kornheiser?) "To write as often as possible, to accept criticism and guidance from someone whose sensibilities you trust and admire, and then to work your butt off. There is no substitute for hard work."

Aw, man. Why can't I just be like a product placement tester? Sit there in a white room all day and try out new gadgets, listen to new albums, view movies while they're still being tweaked? Then blab about why it works or doesn't work, blah, blah, blah . . . And you know what? Forget the white room—they can just send me this shit at home. Why waste all those precious hours commuting?

But no. It looks like (along with every other Joe Blow in this world) I'm gonna have to work for a living.

I open my new spiffy-jiffy Mac email system and see that an email has arrived from David Madden at Fox regarding *Canned Heat*. This is what I'm talking about. I'm about to sell a pilot—that means it's a WGA deal, that right there is gonna be 50 bucks a month added to the trusty pension. But big surprise: My new and improved supernatural redneck version doesn't work for him. The scary stuff feels grafted on. Suddenly *Canned Heat* is neither fish nor fowl.

He's right; I can't argue with him. It felt false from the get-go. I thank Madden for his time and tell him I'll be back with other stuff. I call J.B. and let him know. He shrugs it off. He still hasn't heard from his bro-in-law, David Babcock. J.B. isn't exactly enthusiastic about pursuing the project further. That makes sense. There are only so many chits writers are willing to use up (especially for somebody else's script). So much for *Canned Heat*, for the moment at least. As I've said, this comedy (the unsupernatural version) becomes more relevant each day as the yuppie generation marches inexorably into that purgatory of lost souls and drooping dermatology called retirement.

Then, in perfect timing to allay any doubts, *Point Pleasant* premieres on ABC. Remember? Beautiful, mysterious, witchy, sexy teenager washes up on resort town's beach. And the show tanks. No one gives a shit.

So for now, I lay *Canned Heat* to rest.

I feel myself about to slip into that familiar depressive state which follows rejection. But there's no time for that. I'm agitated, borderline manic. Flipping through the pages of *Variety* when something catches my eye—it seems that ESPN TV is ramping up a daily half-hour show called *ESPN Hollywood*. Wow, this could be right up my alley. I'm the Hollywood Jock, right? So I email the publicist and lay out my background, and he emails back with the contact info for Mike Antinoro, one of several executive producers. A daily half-hour. Do you have any idea how difficult that's gonna be to fill? I can help them here. I go out to the garage and start digging through boxes. Aw shit, not here. So I drive over to my mother's and start digging through her garage. Okay, there it is. *Shaqin' Up*. It's a two-minute video piece I produced at the Hollywood Y, and it's very cool. See, I was hired to find ten black basketball players the size of Shaquille O'Neal to appear in a Reebok commercial. But instead of just renting the small gym at the Y and putting the call out for players, I also brought on my old pal Felix and my former *Blue Chips* assistant Tracey (about the sexiest and smartest P.A. a guy could ask for) to help me shoot a video piece of the process.

The whole thing went great. Felix shot a lot of lively footage, I spent a few long nights creating a paper cut on my VCR, then I hired an editor and rented an Avid Digital editing machine and cut the thing together. And it worked.

I dust off the old VHS, rush home, and plug it in. Okay, it holds up. This is exactly the kind of filler material a show like *ESPN Hollywood* is gonna need. And I'm gonna be the guy to deliver it. Until I make a few calls and realize, Jeez, the guys they're gonna hire for this kinda work are gonna be fresh out of film school and willing to work for peanuts. And their bosses are gonna drive them absolutely crazy in the process. There's an old adage in the world that applies to a lot of things, particularly work: Be careful what you ask for.

So I shove the cassette into a drawer. For posterity. And I get back to the only work that's really exciting me these days, the rewrite of *Zulu Wave*. George, Mark, and I have been really cranking on it. And now it's done. We send it off for some professional reactions. To David Madden at Fox. And to a young guy I met through the column, Steve Shin, who's landed a development job at DreamWorks. And also to a young African-American friend of his, Billy Hawkins, who's working his way up the ladder at Creative Artists Agency. Let's get some feedback here. It's feeling good. It's in nice shape. We'll see what these guys have to say. We know the reactions are gonna be positive. And we're already planning our next step—finding an agent.

48

RUN FROM COVERAGE

Over the weekend, I get an email from Steve Shin (freshly ensconced at DreamWorks). He loves *Zulu Wave*. Feels it's the kind of script he can really champion. He compares it to *Whale Rider* and *Bend It Like Beckham*, and that's not shabby company. Shin asks if he can email the script to a friend of his who does a lot of script coverage for Miramax. This guy will give it a real professional read and do it immediately, as a favor. I say okay, as long as it doesn't go through official channels. The coverage will be strictly off the record.

Well, all right. We're on our way. I drift through the rest of the weekend on a cloud of satisfaction and optimism. *Zulu Wave*—who woulda thunk it? I'm reaching for that third beer when I remember the last discussion I had with Andre about alcohol. Actually, the words were all his:

"Dad, you know why dwunk people aw so dangewous? It's 'cause when they dwink alcohol it makes them happy and they stawt dancing and that fizzes up all the alcohol in theya stomachs so that when they dwive, all the bubbles make them dizzy so they cwash their cahs."

I get to my desk bright and early Monday morning. No word. There is, however, an email from one of the owners of the Summer Pro League. Are we ready to make a deal or not? I consider my options here. Over the years, I've been involved with many things that didn't work. Screenplays, movies, kids' videos, basketball leagues. It's a sick feeling that strikes when the realization finally hits home—this beautiful project is going south fast. It'll either never get launched or it'll slip deeper and deeper into the red. There's always the consideration of lost time and effort. Time you can never get back. I've also lost money on projects. It's hard. And when it comes to other people's money, it's even harder. Regarding the SPL, there are simply too many variables at play to justify ponying up a sizable chunk of cash for this business at this time. I email off our last counteroffer—since they're unwilling to build in the contingencies we require, Clay Moser and I will agree to take over the league and pay them out of profits. It's a win-win situation, I explain. If all their claims are true (and I have no reason to doubt that they're not), in a few short years they'll have pulled even more cash out of the business than they're asking for. The only leap of faith they'll have to take? That Clay and I won't fuck it up. I send off the email. And hear back five minutes later: No way. And if you're the type who reads between the lines that would be: No fucking way.

You know how so many people run around saying, "Hey, it's for the best. It was meant to be that way?" Well, I personally think that's about the dumbest life philosophy one can adopt. But in this case, that old new-age adage is absolutely right. So I lay the Summer Pro League on the shelf, right alongside *Canned Heat* and *China Slam* and *94 Feet* and *Mind Your Own Business* and *Take Me to the River* and God knows what else. Not forever, mind you. Just for now. It's a relief.

But then another bombshell. My book editor at William Morrow, Rob McMahon, calls to say that he's leaving the company. You gotta be kidding me. This is just like the movie business. Rob assures me that my new editor is a sharp young guy who loves my work and everything's gonna be cool. Yeah, right. Then Matthew Guma calls to say the same thing. Aw, man. Is my book gonna die here?

I hang up the phone and shake my head clear. I've got one project that just might pull me through. It's called *Zulu Wave*, and I've got to keep focused. Right after lunch I receive an email from Steve Shin. He's forwarding his Miramax friend's coverage. I eagerly open the document and read this:

> ZULU WAVE tries to blend a ROCKY style underdog story and political drama and in my opinion, fails at both. In regards to the sports movie aspect—I know no more about surfing and competitive surfing (and I know very little) after reading the script than before I started. Interestingly, the same was true of BLUE CRUSH and it was one of its failings. This is a minor note and easily fixed—the impact comes when one realizes that this is symptomatic of a greater problem.
>
> ZULU WAVE is thin.
>
> I will admit that, when cliched emotionally moving scenes presented themselves, I found myself moved. That is the beauty of the language of film and a damn fine reason to steal from successful examples in other films. But in truth, I almost responded to the script by writing: ZULU WAVE—what does that mean? The title alone defines its lack of sense of itself. There is no metaphor there. There is no such thing as a ZULU WAVE and it has no emotional bearing. In telling a fictional story of very real and tragic events, one must (again, in my opinion) honor the reality by thinking deeply about it and having something to say.

I groan. There's more, but I'll spare you the gory details. I quickly decide not to share this coverage with George or Mark or anyone else. Too demoralizing. I fire off an email to Steve Shin—"Coverage like that is why much of Hollywood's best talent leaves in despair. We're determined to put the script into the hands of filmmakers, not readers."

e, you see, is the bane of all those on the creative side of
re's how it works: When you or an agent or a manager send the script to a producer or an executive, the script is often sent off to professional readers for "coverage." Coverage usually runs three to five pages: a couple pages synopsizing the story, a couple discussing merits and problems.

Once a script receives coverage, that's it. It's written in stone. And if it's negative, you're dead in the water at that studio for all time. So coverage (especially early in the game) is to be avoided at all costs because your script is inevitably going to be read by some snide, snot-nosed, overly intellectual, emotionally crippled wannabe screenwriter who instead of putting his balls (ovaries and/or testicles) out there with the rest of us has chosen to sit back and pass judgment, knowing full well that a few choice comments will destroy months of hard work.

"The language of film" my ass.

Now here's where I'm supposed to say: "But you know what? Not all readers are incompetent assholes—some actually understand how to read a script. Some actually know that terse, minimalist dialogue on the page comes to life on the screen. That less is more. That wanting to know what happens next is the surest indicator of a well-written script."

But I'm not gonna do it. You know why? Because script readers are all incompetent assholes. Every last one of them. They're traitors to the cause. Tokyo Roses and Benedict Arnolds. They're the eager kids in the first row with their hands in the air, so determined to make an impression. . . . Only now, here they are, 15 years later, warped by the harsh reality that no one gives a shit about their still-wet-behind-the-ears eagerness. And that it's actually that slouchy kid in the back row who might have a point of view that'll make it to the big screen. So what do these eager beavers do? They become readers, whose main purpose in life is to bury good scripts.

I let it go. I feel the rage, then let it go. We know how good *Zulu Wave* is. Filmmakers have confirmed our belief. Luckily, this guy's coverage was off the record. A favor for a friend. With friends like that . . .

49

BLACK LEATHER, HEAD TO TOE

An email from David Madden arrives: "I liked this script—it delivered on the promise of the outline—only thoughts I'd have would be for the sake of the sale building up a part like Sharkey that you can cast with a Hollywood actor—I know this sounds whorish, but it's going to come up since you obviously want characters like Kwezi to be cast with African unknowns—the other thing is that the tone sometimes seems a little jarring—the politics occasionally feels a little in-your-face for a story whose tone seems closer to something like *Whale Rider* but with more humor. This is a great story and would definitely make a good movie."

Okay, this is auspicious. After everything I've been through with Madden, I'm reasonably certain he wouldn't be bullshitting me about this. I call George and read him the email. Then Mark in New York. They're cautiously optimistic. I email Madden back:

> Thanks, David—
>
> We're with you about the role of Sharkey (although want to hold off beefing up his role until we get the script set up). Also, the role of Jelly is tailor made for Charlize Theron (plus she produces). Regarding the jarring political content—we think the

juxtaposition belongs—it heightens his escape into the ocean. Clearly, it's a director's movie at this point.

So—are you willing to help us? We think Scott Schwartz at Vision Art would be a good agent for this project. He's a surfer and supposedly a good guy. Might you call him for us? 310 888-3288. How about Bob Cort? MTV Films? My book agent in NY is getting it to Dave Matthews' production company (he's South African). And is there a way for Fox to get involved? It's looking like a 5–8 million dollar budget. We'd love a theatrical release (work the festivals) before going to TV.

Best, Rob

Madden emails back. He'd be happy to call Scott Schwartz for us. Meanwhile, I receive word from the young black agent Billy Hawkins at Creative Artists Agency and from Jessica Stamen at Echo Lake Productions. More positive feedback.

I forward Madden's email to both Steve Shin and Billy Hawkins. They're reasonably impressed. But Billy Hawkins isn't saying what I want to hear. He isn't saying, "Let me run up and down the halls of Creative Artists Agency touting this next great script."

And it isn't what George wants to hear either. "I'll tell you what I want to hear," shouts George when I call him. "I want to hear this guy say, 'Okay, I want in. I want to rep this project. I wanna get this fucker made!' That's what I want to hear."

"I'm with you," I shout back. "We're too old for this shit."

George talks about a Danish manager he met at AFI some years ago. "This guy is hot, man. This guy, even way back when he first showed up, you just knew he was goin' places. I'm gonna call him."

"What's his name?" I ask.

"Mikkel Bondesen," George says. "He's Danish. He's this incredible stud Danishman. Back at AFI, he first shows up driving an American muscle car, wearing black leather, head to toe. He says to me [and here George assumes a spot-on Danish accent], 'Georgie, how do you like my outfit? I am quite the stud, huh? My arms look good, my legs look good, and my package, it's all there, am I joking?'"

And suddenly I'm howling. Howling into the phone because I've been flailing away for so many years, and here, somehow, miraculously, we've pulled this beautiful, compelling script out of our collective middle-aged asses, and we need someone to run it around for us, and George is trying to convince me that some leather-bound Dane is just the guy to do it.

"I don't know," I say. "I wanna push through to Scott Schwartz."

"Just Google Bondesen," says George. "Tell me what you find."

So I crank up my new IMac G-5 and Google the Dane but it takes a while because like a lot of geniuses, George's spelling leaves something to be desired—and then suddenly, there he is, Mikkel Bondesen, and he's already got a list of credits and a solid group of writing clients, and I'm thinking, Why not? Everything else is so screwy in the world, why not a mad Dane to run around with *Zulu Wave*?

"Call him," I tell George. "Use the log line: *Zulu Wave*—A young black surfer breaks the chains of apartheid."

"Okay, man, I will, I will."

But it still feels like a long shot compared to surfer-yogi-agent Scott Schwartz. I found him in an article in *Variety* which I promptly tore out and lost in a drawer for a couple of months. He represents Robert Weide, who was a writer-producer for *Curb Your Enthusiasm*. I Googled him, and he turned up in an article about yoga. He was quoted as saying yoga helped his surfing. And my feeble brain went—shazaam!—*Zulu Wave*! (It's hard to be from New Jersey, believe me.)

And now, not only does Madden like the script, he says he'll call Schwartz for us. But this being Hollywood, this simple exchange could take weeks. Months even. So I suck it up and cold-call Vision Art Management and ask for Scott Schwartz. In a moment, I'm on with his assistant, Melissa Cooper.

"What is this regarding?" she asks, and if people like Melissa can sometimes come across as a little snooty, how would you like to field dozens of cold calls from desperate, clueless, brainless, moronic wannabe screenwriters every day of *your* workweek?

I tell her that my partners and I have written a fantastic surfing movie and we're looking for representation, and I say it with conviction, because I believe it. I mean, I'd put *Zulu Wave* up there against *Blue Crush* or *Gidget Goes to Hawaii* anytime.

And before she can say, "I'm sorry, we don't accept unsolicited material," I say, "David Madden over at Fox Television said that Scott would be a great agent for us."

"And . . . ?" she answers.

"And so you ought to read it and recommend it to Scott."

Melissa Cooper says, "Okay, well, we usually don't accept unsolicited scripts, but if you send it, I'll take a look for him." I think—here's a woman who knows her boss. Surfers are passionate people. Imagine if the best script about surfing ever written in the entire history of the universe somehow slipped through his hands, all because of her.

I ask for their email address. "We only accept hard copies. You can put it in the mail."

"Okay, fine, what's the address?"

She tells me, and I say, "And is there an email address where I can give him a brief background on who we are and a log line on the script?"

"Put it in a cover letter," she answers.

Okay, so we're firing on all cylinders. But here comes the latest wrinkle. Not only does *Zulu Wave* go against the grain—too small, too period, and will star an unknown black South African teenager instead of Adam Sandler—*In My Country* is being released this weekend. God, here we go again.

In My Country (starring Samuel Jackson and Juliette Binoche) is a story about the South African Truth and Reconciliation Commission, created to investigate abuses of human rights under apartheid.

Great—it's arriving just in time to muddy the waters for *Zulu Wave*. I go to rottentomatoes.com to check out the early reviews.

"John Boorman's high-minded but hopelessly wooden film makes the fatal mistake of turning characters into mouthpieces."

"Lets no cliché go unspoken in one of the worst scripts of the year."

"Boorman's bathetic tourism is unconscionable." (I'm trying to remember what "bathetic" means—something like pathetic but dirtier?)

"A sludgy quagmire of sanctimonious moralizing and overbearing melodrama." (Hey, I was right!)

"Wholly unsuccessful misfire."

"Earnest but soggy."

"Schlock entertainment."

Oh, God.

I go to www.boxofficemojo.com to check out if they're planning a wide release, which could be deadly. Okay, what a relief—it's a Sony Pictures Classics film, and it's only going out into nine theaters, as opposed to say 1,500 (or 3,500 for that matter).

With any luck, Samuel Jackson will draw a crowd (after all, he was a Wise Man in the Wax Museum crèche). The movie should do just fine per screen (which has gotta be at least $7,000 per) and the good folks at Sony Classics will cut their losses and hope to get their investment back in DVD sales.

Will *In My Country* kill any hope for *Zulu Wave*?

Stay tuned.

50

FABIO WITH A BRAIN

Early Monday morning. I got to www.boxofficemojo.com to see how *In My Country* did. Oh, God. The guys at Sony Pictures Classics must be apoplectic. Here's a movie with Samuel Jackson, one of the hottest actors around, plus Juliette Binoche, and its opening weekend is a grand total of $23,700 from nine screens. That's $2,633 a screen. That's deadly.

But it's a relief in a way, because *In My Country* will disappear faster than *The Last Shot* did a couple of months ago. Out of sight, out of mind. If any executive dares to dismiss *Zulu Wave* because of In My *Country*'s dismal performance, we can rightfully reply, "Wha??!! Never heard of it."

So let's get on with it. This is an important week.

I look for a response from the agent, Scott Schwartz (or Melissa Cooper at least). Nada. With agents, it's always something. Pilot Season. Sundance. Upfronts. Thanksgiving, Christmas. Or writers. That's it. Really, let's face it. It's writers. To agents, writers are just a royal pain in the ass. Why? Because unless they're creating TV shows, there's no real money in writers. Did you catch that *New Yorker* article by Tad Friend about William Morris uber-agent David Wirtschafter? Wirtschafter laid down the law to the junior agents in these words: "Gary Fleder is a director, a very good director, who did *Runaway Jury*. He is not Steven Spielberg, but every time he directs he gets $4.75 million. So you should want to be part of a team developing lots of projects for him. There has to be an allowance for passion, and you may have a hundred-thousand-dollar writer you believe in, and that's great. But it's no longer 'my list, my people, my stuff.' Because you would have to represent forty-seven and a half one-hundred-thousand-dollar journeymen writers to equal one Gary Fleder movie."

So take that.

I call my old producer pal (and former agent) Norman Stephens, tell him we've gone ahead and written *Zulu Wave*. Then I pitch it to him again over the phone. Norman's the real deal. In the trenches. Quietly getting quality movies made (mostly for television, although that's become tougher by the day). Norman is really into the pitch until he realizes it's set in the 1980s. I can hear him go quiet over the phone. It's a death pall. I try to bull through it.

"So whaddaya think?"

"It's period," he says.

"Barely," I answer.

"Look, I don't want to be the naysayer here, but if it happened yesterday or earlier, it's period. You're starting with two strikes."

Now it's my turn to go silent. My mind races with all the exceptions to the rule, but I can't even muster the energy to present them. Norman senses my disappointment. "Look," he says. "You're a friend. I'd love to read it."

We hang up and I email him the script.

Man, it's tough to weather this stuff. It's suddenly occurring to me that I may never hear back from Melissa Cooper. We need representation. We need a savvy champion who's in the middle of the selling-buying game to help get this script set up.

My thoughts return to the crazy Dane, Mikkel Bondesen. Somehow (with George's help) I've built this guy up into mythical proportions—

he's this towering, muscular, long-haired Dane; Fabio with a brain—slicing a path right through Hollywood, irrepressible, indomitable, insane (in the best sense of the word). George breaks through to his assistant, Shannon Pallone at Fuse Entertainment. And she reads the script and gets real excited about it. But Mikkel's wife just gave birth to their second son and Mikkel's pressed, so he reads enough of the script to say, "Okay, Shannon, you want to run with this, go for it."

So now I'm on the phone with Shannon Pallone and she sounds like she's 17 years old or something—highly intelligent, mind you, but still . . . Anyway, despite her youth, she's saying everything I want to hear. Praising the script, how it's one of those that simply has to get made. And I'm thinking, This isn't exactly what I had in mind, I mean I was ready to embrace the Mad Dane, to launch right back into the Hollywood scene—George, Mark, and me—shoulder to leather-clad shoulder, all swinging broadswords, but now suddenly we're getting his assistant, Shannon Pallone.

Shannon Pallone wants to be the point person for *Zulu Wave*. I ask her how long she's been in the business.

"Almost three years," she answers. "I spent a year at Endeavor [a well-regarded agency], then about a year and a half in the indie world. And then I ended up here at Fuse."

"That's cool," I say. "So if you're running with *Zulu Wave*, how do you envision things? What's the strategy?"

"I want to go to Ed Saxon," she answers. "He's a producer. *Philadephia, Adaptation, Something Wild, Silence of the Lambs*.

"Yeah, okay," I say. "I'll look him up."

Shannon continues, "Saxon's affiliated with a company called Participant Productions."

"I haven't heard of them," I say.

"They're pretty new, but they've got the means to finance and they want to make movies with a message and *Zulu Wave* would make the perfect fit."

"Yeah, that sounds good."

"So I'd like to send Ed Saxon the script for a weekend read. Can I do that?"

I think about it. Do we really want such a young, inexperienced person taking the lead here?

"Let me talk to my partners," I answer. "It sounds good, but let me talk to them and call you back in the morning." We tentatively schedule a meeting for next Tuesday afternoon, say good-bye, and hang up.

I IMDb Ed Saxon to see that this guy's produced over a dozen movies, several of which are real gems, including *Miami Blues* with Alec Baldwin and Jennifer Jason Leigh.

If you're looking for a good rental, rent this movie. The limited violence that I've witnessed in my life has been fast, hyperreal and unpredictable, committed by people whom you'd probably feel okay sitting next to on the crosstown bus. That's what *Miami Blues* captures as well as any movie ever made.

Okay, so Saxon is a great choice. Then I track down Participant Productions' Web site, not feeling that optimistic. Participant Productions? What kind of name is that? Really, I mean participants are like the last ones chosen. The spear carriers, the bench warmers. But then I think, Hey, maybe they're so cool, so understated, that the name carries some weight.

I get to the Web site and sure enough, they're the real deal. The founder, Jeff Skoll, cofounded ebay and was their first employee. Do you have any idea how much that is worth in stock options? Me neither. I'll have to ask him if we get that far. Or not. Actually, a quick trip to Google delivers this from an old *Fortune* profile of rich people under 40: Jeff Skoll—Current Net Worth ($): 3.85 billion.

God, this guy is richer than Mark Cuban. Remember, one billion dollars is one thousand million dollars. One thousand million. Not ten million, not a hundred million, A THOUSAND MILLION!!!

Back to Participant Productions. This is from their Web site—"Participant exists to tell compelling, entertaining stories that also create awareness of the real issues that shape our lives."

Cool. They've got an impressive list of executives and a growing list of projects either in the can, shooting, or on their way, including the "Untitled Niki Caro Project" starring Charlize Theron. Niki Caro got an Oscar nomination for directing *Whale Rider*, a film we tout as a soul sister to *Zulu Wave*.

Then, going through their executive bios, I discover that Chris Adams, chief vision officer (whatever that means) and senior VP of business development, is an avid surfer.

Aha. (Not that the surfer connection is doing us much good with Scott Schwartz.)

I immediately email Shannon Pallone at Fuse, telling her what I've found. I also ask if she went to Duke (a Google turned up her picture in some friend's blog). She emails back that she is indeed a Dukie and I think, Uh-oh, my book agent, Matthew Guma, went to Carolina and we're heading into March Madness. This could be trouble.

But I call her and say, "Sure, send the script to Ed Saxon's person for a weekend read. As long as we get a response by Monday."

"Great," she answers. "Have a good weekend."

51

THE SKELETON WITH A MARGARITA

Monday morning, I'm waiting for a response from Ed Saxon's guy when this email arrives from producer Norman Stephens (the same Norman Stephens who worried that the "period" aspect of *Zulu* would make it much harder to get made).

> Dear Rob—
>
> I think your script is terrific and would like to share it with David Skinner for consideration as a ShadowCatcher project. It's a great "outsider" sports movie with the strong overtones of racism. I think it transcends the period.
>
> Best Regards,
>
> Norman

Okay, cool. That's cool. And you gotta love that phrase, "transcends the period." I make a mental note to trot that out the next time "period" becomes an issue. But who the hell is ShadowCatcher? I Google 'em up.

David Skinner is old Seattle money. He's produced several credible indie films—most notably *Smoke Signals,* which really worked for me. Since then, they've made several others, including one recently screened at Sundance, *Game 6*, starring Michael Keaton and Robert Downey Jr. (written by the brilliant Don DeLillo).

All right, cool. It feels like a long shot, but you never know.

I next call Franklin Leonard at Leonardo DiCaprio's Appian Way. Franklin's the guy George and I pitched to at John Goldwyn's at Paramount. He's since seen the writing on the wall and moved onto a sunnier work scene. Unfortunately, when I Google Appian Way, I discover that they have another surfing movie in the works—the story of Miki Dora, one of the original Malibu surfers in the '50s and '60s, and Leonardo is set to play the lead.

Nonetheless, Franklin returns my call. Suggests Will Smith's Overbrook as a good home. Also, Oprah's Harpo Films.

So over the weekend, North Carolina beat Duke. I figure I'll get a good sense of my possible new manager's commitment to college hoops by her response. I send her this email:

"I thought after yesterday's game you might've been too depressed to show up for work today. Jeez, Duke with a six point lead with 1:30 to play . . ." And minutes later, I get this reply from her:

"I totally considered calling in, but there's far too much to be done. And at least it was a close game! And I was happy to see OSU and Mizzou win."

Shannon Pallone! Who knew you were such a hoops nut? This should bode well for my college hoops movie, *94 Feet*. But I catch myself. The last thing I want to do right now is muddy the waters. See, if I make a run at Shannon with other scripts, I know that George and Mark will start yanking stuff off their shelves to throw at her, and before you know it she'll be knee-deep in soggy, dog-eared, blood-encrusted screenplays.

Keep your eye on the prize. All our time and energy must go to putting *Zulu Wave* in the proper hands with the right hype.

Tuesday morning, I pick up George and we do the trip into Hollywood to meet Shannon Pallone face-to-face and decide if she should be the one to manage us.

Fuse Entertainment has their office on "The Lot," which used to be the old Warner Bros. studio before they moved to the Valley. Formosa and Santa Monica Boulevard. I remember that almost a year ago, I was at this exact same location, staggering around the Reebok/Baron Davis NBA All-Star party, trying to pick a fight with Dennis Rodman. Now I'm back to sell a script about a Zulu surfer.

We pull into The Lot in my old Camry. Exactly nine months ago, George and I were pulling onto the Paramount lot to pitch *Zulu Wave* to John Goldwyn's development guy. There the security was so tight I'm surprised they didn't find the lump developing in . . . (whoops, I'm getting ahead of myself). Anyway, here at The Lot, things are a lot laxer. Then again, if some al Quaeda lunatic was looking to blow up a Hollywood studio, chances are he wouldn't pick The Lot. This is the kind of working-class movie lot where the World Poker Tour is rubbing shoulders with the Mad Dane.

But the only way to park is valet. Sheeesh, there goes another two bucks. And in a misguided effort at boosting your karma, you know you gotta make it five.

I tell the valet as he climbs into the Camry, "Don't scratch it, okay?"

He laughs.

Suddenly a brand-new BMW pulls up. I turn to see George eyeballing the two young guys climbing out of the Beemer. He's snarling almost. These two youngsters are clearly writers—T-shirts, baggy jeans, black Converse low-tops. And doing well for themselves—with that phony nonchalance that's making up for all the snubs they received in junior high.

I feel my lip curl. George and I could pound the shit out of these two guys in about ten seconds. One guy in particular is coming on a little thick with the "Ain't I cool?" shit and I'm thinking, Hey, asshole?! Yeah, you, hot shit with the lucky early break, I was wearing black Converse low-tops when you were sucking your toes in your mother's womb, you lucky prick, so don't you . . .

But then I let it go. It's like I'm doing a shitty imitation of Denis Leary in *Rescue Me* and I know I gotta let it go. Hey, I had my shots. Way back when. I had my share. I scored some points. Got my share of deals. Got vested in the WGA. Got a chance to write this book.

And now I got another shot. *Zulu Wave.*

Keep your eye on the prize.

George and I enter the Formosa Building, walk down a long shabby corridor, climb a shabby staircase (we're talking chipped paint, worn carpet), and wander into the offices of Fuse Entertainment.

Ryan's the guy manning the front desk (I'm glad it's not Shannon). Ryan's got a headset on, pounding out an email. We shake hands, introduce ourselves. He tells us Shannon will be a few minutes. George and I sit and chat him up. Turns out he's read *Zulu Wave* and likes it. So that's cool. But then again, what's he gonna say if he thought it sucked—"I thought it really sucked"?

Shannon Pallone appears in the doorway and says hi. This young woman comes as no surprise. To me at least—'cause remember, I Googled her and turned up a picture. Plus we've been emailing back and forth. Unlike her brilliant, hyperkinetic boss (whom we still haven't met), she's chosen to go the low-key route. She's pretty and pleasant and highly intelligent.

George and I chat her up. God knows what she thinks of us. Between us, we're 103 years old—probably four times her age. But we give a shit. We're passionate about *Zulu Wave*. And we let her know it.

She shares the sentiment.

Again, we return to our theme. A few select scripts absolutely deserve to get made, for all the right reasons (in exactly this order): One, they'll make money. Two, they'll bring prestige. And three, they'll help make this fucked-up world a better place for our children.

Then George and I pepper her with questions.

Who else is in the company? (Alex Goldstone, Mikkel's partner, plus they'll be adding more managers and a production exec soon.) How much movie work do they do as opposed to TV? (It's a mix.) How often do they attach themselves as producers to projects their clients generate? (Not that often.)

Shannon goes on to say how they like to deve
career strategy with their writers. Five years. Ten years.

And I'm thinking, Ten years? What are you crazy? In ten yea
gonna be on a beach someplace—either with a margarita in my hand,
or dead, or both, take your pick. And fuck even five years! Here's what we need to know—how are you gonna help turn *Zulu Wave* from a script into a movie?

Because that's what it's all about. All those scripts, yearning to be movies. They're like frozen embryos. Stashed away everywhere you turn. Open up any freezer door: a blast of fog and cold air—and the yearning for life.

Shannon Pallone, will you be the birth mother of *Zulu Wave*?

Somehow, I keep myself from asking her.

Instead, I tell her I'm picking Carolina over Duke in the semis, but Illinois to win the whole thing. She tells me she won't hold it against me 'cause Duke's goin' all the way.

George gets us back on track. Willing to ask the hard questions. Because Shannon Pallone is still a young pup in a world where relationships (between sellers and buyers) are everything. How much will Mikkel participate in this process?

"Every step of the way," she assures us.

Her agreement with Mikkel is this—she'll continue to assist him until she's generating enough income as a fledgling manager for Fuse to cover her salary, at which point she won't have to answer his phones anymore. Fair enough.

We discuss strategy. The script should go out to a handful of producers. I pull a single sheet of names from the folder I'm carrying. A list of 35 or so. They range from Will Smith to Oprah Winfrey, Scott Rudin to Ed Saxon.

These sorts of lists are easily constructed. If you're interested in playing the screenwriting game, I recommend devoting a few hours a week to reading the trades, poring through the latest deals, acquainting yourself with who's buying, who's selling, who's producing, who's directing. Thumb through the Hollywood

Creative Directory, the online sites. Get a handle on the flow of the town.

And if you're one of those arrogant assholes who refuses to move to Hollywood —give it up. Or write a book. Or better yet a play. And put it on with the Brattleboro Players.

Screenwriting is a game for those willing to pay the dues. (But believe me, if paying dues for a single twenty-something is to move to L.A. and flail away with a bunch of other footloose-against-the-grain types who understand that the most bang for its buck when it comes to entertainment is free sex, then what's the hardship?)

But be forewarned. If you don't believe in THE VALUE OF PROCESS, you're gonna get burned.

Screenwriting is really analogous to March Madness. Only here, you get to write your own invite by simply moving to Hollywood. Do it, and guess what? Suddenly you're in the dance.

But remember, each March, as 65 teams vie for the championship, 64 will finish with a loss. Not as losers, but with a loss.

The trick about being successful in both sports and Hollywood is this—the moment you make yourself a player, you've won. Otherwise, you're bound to be a loser.

We wrap up with Shannon Pallone by telling her we've also shown the script to John Hayes at Cube-Vision and Couper Samuelson at 2929 Productions. She says she'd like to send it to a friend at Tollin/Robbins. Cool by us. These early submissions will tell us a lot. Then we shake hands and we're out.

We've got a manager.

52

A FROZEN BAG OF PEAS

So now it's a waiting game. And the waiting is killing me. I call George.

"Hey," I say.

"Hey, whatcha up to?"

"I'm sittin' here with my thumb up my ass," I answer.

"Me too," says George. "But you know what? It feels kinda good. And you know why? 'Cause yesterday, I swear to God, I thought I was gonna die. I thought I was dead."

"How come?" I ask.

"Because I woke up and I found a lump in my scrotum."

"That's your other ball," I say.

"No, I swear to God. I fucking freaked out. It was cancer. It had to be. I was dead. I was writing my goddamn obituary. So I called out to Liz, I got cancer! I'm dead! And you know what she says to me? 'You're not dying. You've got a lump in your scrotum, that's all.'"

And here I should say that Liz is a cancer survivor so she knows a thing or two.

George rants on, "But do I buy that? A lump in my scrotum . . . like what, it's some stray Milk Dud or something? No fucking way! I am dying, no doubt about it. So we get to the doctor, and he looks at me and laughs. He laughs and says, 'You're a writer, right? So you got what eighty-five percent of my writer patients get—nonbacterial epididymitis.' And Liz says, 'See, I told you.' And the doc says, 'Look, you spend all day sitting on your balls, right? That's what writers do. And your balls get tired of it and develop this lump. You gotta understand, you writers are in just about the most dangerous profession out there. Forget firefighters, forget steelworkers. Writing's the worst—high stress, constant rejection, alcoholism, drug abuse, heart disease, phlebitis, self-contempt, you name it, plus you sit around on your balls all day.'"

George rants on—"So by now I gotta say, I am like so elated. I'm fucking beaming! I'm gonna live! I'm gonna live! I'm serious, I really thought I had cancer. And the doctor says, 'Look, all you gotta do is go home and get a bag of frozen peas and stick it under your balls.'"

"'You want me to put frozen peas under my balls?'"

"'Yeah. You can leave them in the bag, though. So go on, get outta here. Go home and elevate your balls.'"

"So Liz and I are walking out through the waiting room—you know how sometimes you're so happy you wanna kiss a total stranger? So there are all these people sitting there, and I look over at these three old Armenian ladies—I'm talking babushkas, black dresses, baggy black stockings—and I walk up to them and announce, 'I am so happy right now. You know why? Because 20 minutes ago I walked in here convinced I had cancer, and now I'm walking out, and you know what the doctor told me?! The doctor told me what I've got is nothing, it'll go away, just go home and get a bag of frozen peas and stick it under my balls!' And everyone is like freaked, right? Like who is this lunatic? But then one of the old Armenian women looks at me and in this really fractured accent she says, "'So God bless for you and your balls!'"

"And we fucking howled! The whole waiting room was like on the floor peeing in our pants, man!"

George.

Norman Stephens calls to say that David Skinner and Tom Gorai of ShadowCatcher Entertainment love *Zulu Wave* and Skinner wants to know more of its history before making an offer.

I give Norman an honest appraisal. It's brand-new. We've tested the waters with a few companies and are getting a terrific response.

"Would you be willing to have David take it off the market?"

"What are the chances of you guys actually making the movie?" I ask.

"They're good, Rob," answers Norman. "Listen, David Skinner is the kind of guy who doesn't snatch up material unless he's serious about financing it. He likes your script. He thinks it's a terrific story. He loves the fact that it's set in the midst of the conflict over apartheid; he thinks that's one of its strengths. He may go looking for an international partner to put up half the financing, but he's interested in keeping the North American rights."

"It can find a substantial audience," I tell Norman.

"We know that. That's why he's interested. That's why I'm interested."

"We get first rewrite," I say.

"Of course," Norman answers.

"We don't want to be producers," I say.

"How about your management company?"

"Same thing. They just want their commission and to help any way they can."

"Good," says Norman. "And remember, I'm going to be producing it. I'll keep you guys in the loop as much as possible."

"So what kind of money are we talking here?" I ask.

"It's going to be a modest offer—10 to 20 thousand to option it. A guaranteed WGA rewrite. A decent low six figure back end. It'll be a fair offer for this kind of indie picture."

"Okay, let me talk to my guys," I say.

I hang up. Holy shit. We just got a deal.

A flurry of activity. Emails, phone calls. ShadowCatcher's lawyer will be emailing their offer to Shannon Pallone for us later that day. But should we accept it? Why wouldn't we? Here's why—because it's so early in the game. We know we've got a valuable script. Should we take it off the market just like that? Jeez, what if Tollin/Robbins calls tomorrow saying they love it?

We instruct Shannon to acknowledge their offer and tell them it'll be a couple days for us to get back to them. In the meantime, Shannon recruits Mikkel and they start putting on pressure for a response from the few companies the script has gone to. Plus Mikkel wants to send it to one other guy. A guy who also finances movies, Ben Goldhirsh.

Okay, fine, good. I email Norman that everything is on track. Then I Google Ben Goldhirsh and find his one and only IMDb.com credit: *Jakub Zvejkal Is Les Boogie* in 2001. The tag line—The first step toward realizing your dreams is to wake up.

Hmmm. I'll have to sleep on that one.

But this is feeling good here. And for the next 48 hours we hang tight—reviewing the offer, preparing a counter, and waiting for other responses to trickle in.

Couper Samuelson at Cuban's 2929 passes. While he liked the writing, they're looking to make bigger movies. This one always gets me. "Bigger movies." Like what, *Titanic*? *Star Wars*? What if Steven Spielberg wanted to direct *Zulu Wave*? Suddenly it's a big movie. For God's sake, what was *The Terminal* before Spielberg and Tom Hanks committed?

I feel like writing Couper back at 2929 saying, "You mean you want to continue making films like *The Jacket* starring Adrien Brody which just opened on 1,331 theaters for a paltry $2,406 per screen? Big movies like that, you mean?"

Too small my ass. (Or maybe *The Jacket* is why 2929 is changing its m.o.)

You've got to expect that most companies are going to pass. You gotta brace yourself. You deal with the rejection by turning it on them. Morons.

But the pressure mounts. Then, in quick succession:

Tollin/Robbins passes—they've got an African story.

Cube-Vision passes—no reason.

Participant passes—too small.

Echo Lake is still on the fence.

Ben Goldhirsh passes—no reason.

And finally I hear from Guma—Dave Matthews's company, ATO, passes. Matthews doesn't want to be too compartmentalized with South African films.

Uh-oh. Suddenly the writing is on the wall. *Zulu Wave* is not gonna be that easy a sale. We tell Shannon to call Norman. Mark Rogers, George Walczak, and Rob Ryder agree to take *Zulu Wave* off the market while we hammer out a deal with ShadowCatcher Entertainment.

53

BABY, LET ME FOLLOW YOU DOWN

Look, I'll be honest. I didn't actually pull this off in 52 weeks. My lovely wife gave me a six-week extension. But even that extension had to be extended. Because from the moment we verbally closed the ShadowCatcher deal to when we could actually cash the check was over two months. Usually this stuff lasts forever.

A lot happened during this time.

Dennis Miller's show got canceled.

Ice Cube's *Are We There Yet?* ended up grossing over 60 million domestic.

Oren Koules never called me back.

I implemented David Allen's organizational system, "Getting Things Done," and it's changing my life.

Pat Tillman's parents accused the military of lying and disrespecting his memory for propaganda purposes.

American troops continued to die in Iraq.

While Iraq started getting cozy with that other axis of evil, Iran. (For this we've sacrificed how much?)

A study found that teens pledging virginity until marriage are more likely to have oral and anal sex than other teens.

Nigel Miguel and his partners started prep on their third low-budget movie in 12 months

Pookey Wigington secured financing for his own comedy club.

Ron Shelton got another movie going.

Ben Affleck dropped out of *Glory Road.*

John Carls had a movie put into production.

Uber–sports agent Aaron Goodwin was fired by LeBron James.

Andre lost his two front teeth.

Cole had his braces removed.

National Geographic Feature Films came on board with ShadowCatcher to produce *Zulu Wave.* (And this is truly serendipitous, since their movie *March of the Penguins* has raked in over $75 million domestic.)

Jeremy Cesarec, my new editor at William Morrow, turned out to be a stand-up guy and it looks like the book's gonna be published.

My wife didn't leave me.

And I decided it was time to go see a shrink (God pity this poor soul).

Finally the check arrives in the mail. But it's made out to all three of us. Aw, man. Mark the travel writer is down in the Amazon or something. Do we dare return the check and run the risk of having them go through the whole thing again? Run the risk of a sudden case of buyer's remorse where ShadowCatcher suddenly backs out and disappears?

No way. Not us. No, George and I do what any right-thinking, supernatural rednecks from New Jersey would do—we forge Mark's signature on the check and march it into the nearest bank.

Twenty-four hours later the check clears, and we're each $5,000 richer. That's about $3,500 after taxes.

Like it's been said: Intermittent reinforcement is a dangerous thing.

But there's a paid rewrite in the offing. And the promise of more work based on this sale. And now all my other stuff will be suddenly worthy of another look from the powers that be.

Time to dust off the old raincoat. Enough of this intermittent reinforecement horseshit. Give this Hollywood Jock at least fifteen minutes of satisfaction here. Time enough for George, Mark, and me to sit back, put our feet up, clink beers, and recount once again the words of that old Armenian babushka, "So God bless for you and your balls!"